WORLD CATALOG OF GENUS-GROUP NAMES
OF BEE FLIES

NEAL L. EVENHUIS is an Associate Entomologist with the J. Linsley Gressitt Center for Research in Entomology, Bishop Museum.

World Catalog of Genus-group Names of Bee Flies (Diptera: Bombyliidae)

NEAL L. EVENHUIS

Bishop Museum Bulletin in Entomology 5

Bishop Museum Press
Honolulu, 1991

PUBLISHED BY
Bishop Museum, Honolulu
W. Donald Duckworth, Director

Printed in the United States of America
Date of Publication: 15 September 1991

The paper used in this publication meets the minimum requirements of American National Standards for Information Sciences—Permanence of Paper for Printed Library Materials
ANSI Z39.48—1984

ISBN 0–930897–56–0
ISSN 0893–3146
Library of Congress Catalog Card No. 91–072826

Contents

THIS WORK BENEFITTED greatly from review by Jack C. Hall, David J. Greathead, and F. Christian Thompson. Special thanks are due to David K. Yeates for kindly providing information from his research on genera within the tribe Comptosiini including synonymies and pointing out the lack of pile on the metapleuron in *Anastoechus*. Curtis W. Sabrosky generously supplied corrections and information from his notes. Pam Gilbert was especially helpful with the arduous task of dating many of the more obscure articles. The librarians and staff of the Bishop Museum and Smithsonian Institution are thanked for allowing me access to publications in their collections, which aided the publication dating process of many of the works in this study. Jon Gelhaus and David Yeates provided help with dating difficult-to-find articles after this work had been submitted for publication.

Abstract

A CATALOG IS GIVEN of all the genus-group names proposed for the family Bombyliidae up to 1990. This paper presents a comprehensive overview of the nomenclature of the Bombyliidae at the generic level and stabilizes generic concepts based on these names. Since 1758, 451 genus-group names have been proposed including 24 fossil and amber genus-group names, of which 278 are considered here as taxonomically valid. All names are listed alphabetically and the following information is given: type species given with mode of type-species fixation, proposed positions in the classification within the Bombyliidae, and currently accepted junior synonyms. A current classification of the family is presented, which is intended to serve as a foundation for ongoing studies on the higher classification of this and related families of flies. Twenty-one new junior synonyms for genus-group names are proposed: *Argyramoeba* Loew = *Spogostylum* Macquart, **n. syn.**; *Astropha* Rafinesque = *Cytherea* Latreille, **n. syn.**; *Cladella* Hall & Evenhuis = *Mythenteles* Hall & Evenhuis, **n. syn.**; *Dasypalpoides* Evenhuis = *Dasypalpus* Macquart, **n. syn.**; *Eccremus* Wiedemann = *Tmemophlebia* Evenhuis, **n. syn.**; *Eucharimyia* Bigot = *Euchariomyia* Bigot, **n. syn.**; *Litorrhynchus* Verrall = *Exoprosopa* Macquart, **n. syn.**; *Meganedys* Hull = *Aphoebantus* Loew, **n. syn.**; *Nettadion* Gistl = *Lomatia* Meigen, **n. syn.**; *Nevria* Agassiz = *Comptosia* Macquart, **n. syn.**; *Pioperna* Enderlein = *Systropus* Wiedemann, **n. syn.**; *Proglabellula* Hennig = *Glabellula* Bezzi, **n. syn.**; *Ptinia* Rafinesque = *Phthiria* Meigen, **n. syn.**; *Protepacmus* Cockerell = *Villa* Lioy, **n. syn.**; *Protolomatia* Cockerell = *Paracosmus* Osten Sacken, **n. syn.**; *Stygides* Latreille = *Lomatia* Meigen, **n. syn.**; *Stygidium* Berthold = *Lomatia* Meigen, **n. syn.**; *Subbombylius* Kapoor = *Amictus* Wiedemann, **n. syn.**; *Tauropsis* Hull = *Exhyalanthrax* Becker, **n. syn.**; *Tornotes* Gistl = *Conophorus* Meigen, **n. syn.**; *Zerinthia* Rafinesque = *Usia* Latreille, **n. syn.** One new species-group synonymy is proposed: *Mythicomyia mutabilis* Melander = *Mythicomyia propleuralis* Melander, **n. syn.** One type species is designated herein: *Hyperusia minor* Bezzi, 1921 for the genus *Hyperusia* Bezzi, 1921, **pres. des.** New generic status is given to *Mythenteles* Hall & Evenhuis, **n. stat.**

Introduction

BEE FLIES (BOMBYLIIDAE) CONSTITUTE one of the largest groups within the orthorrhaphous Brachycera with over 5,500 species-group names having been proposed since 1758. They are found in arid to semi-arid area, primarily in the temperate latitudes of the globe, but some genera can also be found in more tropical portions and even some oceanic islands (though rare on the latter). The family is distributed worldwide except in the polar regions. Adults are predominantly heliophilic and many of the larger representatives are commonly found basking in the sunshine on bare ground, rocks, or twigs, with their wings outstretched. All, except those with vestigial mouthparts, are nectar and pollen feeders and are important factors in the fertilization of many desert annual wildflowers. Knowledge of the immature stages is modest because of the relative rarity of encountering immatures in collecting. Published records, however, indicate that most are either predaceous, inquilinous, or parasitic on the eggs and immatures of a variety of orders of insects that dwell in soil (i.e., either underground or in mud cells). Some potential as agents of biological control has been considered in some groups that are parasitic or predaceous on economically important insects (viz., *Exhyalanthrax* spp. on *Glossina; Systoechus* spp. on Acrididae).

Recent research has indicated that the Bombyliidae are undoubtedly paraphyletic and probably even polyphyletic (Mühlenberg 1971; Hennig 1973; Theodor 1983; Bowden 1985; Evenhuis 1990a), which is not surprising given that the family has never been properly defined and has served as a "dumping ground" for a wide variety of disparate groups that just "fit best" in the Bombyliidae. Historically, the family has been split into 2 groups. Initially, these groups were the Bombyliidae and Anthracidae of Fallén, Zetterstedt, Leach, Zeller, Latreille, and Macquart in the early 1800s. Meigen (1820) united these 2 families into a broader concept of the single family Bombyliidae. A century later, after the family had grown tremendously in numbers of genera and species, Bezzi (1924) proposed dividing it again, but this time into "sections" (a level between family and subfamily). These were namely the Homoeophthalmae (Bombyliidae-like taxa with the hind margin of the eye not indented and without a bisecting line) and the Tomophthalmae (Anthracidae-like taxa with the hind margin of the eye indented and with a bisecting line). Mühlenberg (1971) showed that only the Tomophthalmae was monophyletic and left the

Homoeophthalmae open to further research. Rohdendorf (1964) proposed further dividing the family into 4 separate families (Bombyliidae, Cyrtosiidae, Usiidae, and Systropodidae). Rohdendorf's (1964) lead in separating the last of these families was followed by Hall and Evenhuis (1980) and Evenhuis (1982), but recent unpublished phylogenetic evidence indicates that this action was unwarranted and that those taxa comprising that group should be considered as the subfamily Systropodinae. Evenhuis (1990a) has recently shown that family status for the Usiidae is also unwarranted; however, there is recent unpublished evidence that the smaller representatives of the Bombyliidae (microbombyliids) might merit family status as the Mythicomyiidae (= Cyrtosiidae of Rohdendorf). A complete phylogenetic analysis of the family, which considers the intra-subfamilial relationships as well as relationships among other families of Asilomorpha, may well show that the original division of this family into Bombyliidae and Anthracidae needs to be reconsidered. Until then, all the elements of these heterogenous groups of taxa are provisionally considered within a broad concept of "Bombyliidae."

Catalog Rationale and Notes on Hull's (1973) Work

This is not the first compilation of genus-group names of the Bombyliidae. Hull's (1973) work was the first to compile a large amount of the taxonomic and biological information into 1 volume. Its usefulness as a reference for all bombyliid workers cannot be overstated. However, there are many inherent problems such as errors in citation of authors, dates, pagination, and even spelling of genus- and species-group names, which can easily be perpetuated by the uninitiated reader. Compounding this is the unfortunate fact that many new taxonomic and nomenclatural actions are either done cryptically or with a lack of regard for nomenclatural rules. For instance, in his list of genera on page 61, there is "*Paramonovella*, new," but there is no mention anywhere else in the text of this name. On page 445, Hull wished to propose a genus-group name for a group of species with similar habits to *Walkeromyia* species. In proposing the name *Stymphalina* he made 2 errors: first, he called *Stymphalina* a "new name." It is not a new name in the strict sense of the term because it is not replacing any other preoccupied name. Second, and most important, he gave no characters that satisfy the criteria in the International Commission of Zoological Nomenclature's (henceforth I.C.Z.N.) *Code of Zoological Nomenclature* (I.C.Z.N. 1985) to make *Stymphalina* an available genus-group name; hence, it must remain a *nomen nudum*. Other cases, such as *Chelina*, require a bit of research and interpretation of the I.C.Z.N. Code to discover that there is actually enough information presented in Hull's work to make available both the generic name *Chelina* and its type species. There are too many corrections to the entire work to be printed here (they are the subject of another work in preparation), but suffice it to say that one must use Hull's work with much caution.

This catalog has accumulated all of Hull's genus-group names (valid and invalid) and their multiple spellings and clarifies the status and type species of each. Hull's (1973) work had a cut-off date of 1965 (though he did include some published names up to 1971). In the intervening 29 years there has been considerable taxonomic work done on the family and many new genera have been proposed. Evenhuis (1983b) provided a bibliography and list of genera and species known till then, but did not provide information on type species, nor was there any attempt made to identify all the genus-group names, including emendations, that had been proposed for the Bombyliidae.

This catalog is the result of many years of research into nomenclators, catalogs, and taxonomic works to find all the genus-group names attributed to Bombyliidae and present them in a single place with a list of currently recognized synonyms. A total of 451 genus-group names are listed in this catalog, including 24 fossil and amber genus-group names. Of the 451 names, 278 are currently considered as taxonomically valid. Some 21 new synonymies of genus-group names are proposed herein, based either on objective synonymy or examination of type material. One subgenus is given generic status herein.

A concerted effort was made to ensure accuracy with regard to citation, dating, and spelling so that this catalog could be used by subsequent researchers on Bombyliidae without having to consult a diffuse array of references needed to verify correctness of a citation. Despite all the best efforts, errors may still appear. The reader is encouraged to let the author know of any that are found so that they may be entered into an ongoing database of these genus-group names.

Catalog Format and Explanation

All genus-group names proposed since 1758 (available and unavailable) are listed in alphabetical order, including names based on fossil and amber specimens, *nomina nuda, nomina dubia*, emendations, and multiple original spellings. Subsequent misspellings listed in nomenclators have been entered with cross-reference to the name under which each is discussed. Valid genus-group names are printed in boldface; invalid and unavailable names (junior synonyms, junior homonyms, unjustified emendations, multiple original spellings, subsequent misspellings, and *nomina nuda*) are in italics; vernacular names (e.g., those in Latreille [1825]) are in plain roman. Unavailable names are distinguished by a colon between the genus-group name and the author. Genus-group names whose identity is uncertain through vague descriptions and/or loss or unavailability of type material are labelled as *nomina dubia* but listed as valid until their status can be clarified. Genus-group names based on fossils and amber inclusions are denoted by a dagger (†).

Authors' names are attributed to genus-group names according to the I.C.Z.N. Code Art. 50.

The year of publication corresponds to the dating given in the Literature Cited. In some cases, the year is different than that given in previous works. Research was undertaken to ensure that the most accurate date possible was obtained for each work cited in this catalog. The day and month or month of actual publication is placed in parentheses after the bibliographic citation in the Literature Cited. When no mark or annotation is given, the date given is the actual publication date printed on the original publication or that was found somewhere in the journal or book concerned. Library receipt dates were used to provide earlier dating than was given in the original work and are marked by notations to the dates. Dating notations are as follows: single asterisk (*) indicates date recorded as received at The Natural History Museum (BMNH), London; double asterisk (**) indicates date recorded as received at the Bishop Museum, Honolulu. When no further date could be ascertained except for the year of publication, the last day of the year is given and marked with a plus (+) to distinguish from actual publication dates of 31 December. In cases where other evidence (published and unpublished) was used to ascertain publication dates, these data are explained in annotations under each reference concerned.

Cases where an author describes a genus-group name as new in more than one publication have each citation mentioned with subsequent citations in brackets. Reprinted papers, such as those for Macquart, Bigot, Loew, and Rondani, are also listed because the reprinted version is often more accessible than the original journal version. In the case of Macquart, the format for listing reprinted versions follows Smith et al. (1980) and Evenhuis et al. (1989) by listing pagination in parentheses except for difference of year or when a reprinted version appeared earlier than a journal version.

The page number given for each genus-group name is in some cases different than that listed in previous works. The original publication for each genus-group name was searched and all pages referring to it noted. Those listed here are for (1) available names (the pages(s) necessary to make the name available) and (2) unavailable names (the 1st included mention of that name). The listing of a new genus-group name in a key on an earlier page than the description and inclusion of species is often overlooked by catalogers and taxonomists. However, in some cases, such a listing can change not only the year of the validation of the genus-group name but also the mode of type species fixation (e.g., see *Epacmus* Osten Sacken). Unless noted otherwise, all names were proposed as genera.

For all available names, the type species is listed in its original orthography given by the author designating the type species. Type species are also given for *nomina nuda* and other unavailable names to fix their synonymy in a database of generic names of Diptera (some of these are new designations but have no status in nomenclature). If the type species has been found to be synonymous or homonymous with another species, additional clarification of the currently accepted type species is given in brackets (e.g., "Type species: *Ogilviella tridentata* Paramonov, 1954 [= *Isotamia daveyi* Bezzi, 1912]" and "Type species: *Thlipsogaster ater* Coquillett, 1894 [preoccupied, = *Bombylius coquilletti* Williston, 1901]"). Mode of fixation of the type species follows the type species information. These include: monotypy, original designation, subsequent designation, subsequent monotypy, and objective synonymy (here termed simply "automatic"). There is 1 case of a type species being designated by prior monotypy *(Coquillettia)*. Definitions of most of these terms can be found in Blackwelder (1967) and I.C.Z.N. (1985). There are no cases in Bombyliidae of type species by tautonymy. In those cases where mode of fixation has been listed as by original designation and monotypy (e.g., Bowden 1980), the original publication citing each name has been checked for mode of fixation. All were found to be fixed by monotypy according to the changes in the most recent I.C.Z.N. Code. Cases of a genus-group name based on misidentified type species should be presented to the I.C.Z.N. Commission for plenary action. Five such cases exist in the Bombyliidae.

Reference to the 1st included synonymizer for junior synonyms is given after the type species information. Synonymy of unavailable names is given when known. New synonymies for available names are given in boldface. If a genus-group name is based on fossil or amber specimens, this is specified. The current classification of all valid names is given in boldface according to the classification listed below. Subgenera are listed in boldface as a cross-reference with the latest publication source for treatment of the subgenera (where applicable).

A remarks section clarifies certain taxonomic or nomenclatural circumstances, for instance when multiple original spellings need revision or errors presented in previous catalogs and nomenclators need correction.

References Consulted

An initial list of genus-group names of Bombyliidae was obtained from Hull (1973) many years ago and was used as the foundation for this work. Subsequent publications were consulted and new names continually added to this listing. For this specific project, all available nomenclators were scanned for all listed names attributed to Bombyliidae. In addition, larger works such as Hull (1973), Wiedemann (1828, 1830), Fabricius (1805), Engel (1932–1937), Bezzi (1921a, 1924), Paramonov (1940), Hall (1976a), and many others, were read page by page to search for all names proposed in Bombyliidae, including those in footnotes. A few names not recorded in nomenclators or regional Diptera catalogs were found in this fashion.

The following is a selected list of regional catalogs, nomenclators, and major taxonomic works consulted during the preparation of this catalog:

Nomenclators. Agassiz (1846, 1847); Edwards and Hopwood (1966); Edwards and Vevers (1975); Marschall (1873); Neave (1939a,b, 1940a,b, 1950); Schulze, Kükenthal, and Heider (1926); Schulze et al. (1927–1929, 1929–1932, 1932–1935, 1936–1954); Scudder (1882a); Sherborn (1902, 1922–1932, 1932–1933).

Regional and World Catalogs. Aldrich (1905); Bezzi (1903); Bowden (1975c, 1980); Evenhuis (1989); Kertész (1909); Painter and Painter (1965); Painter, Painter and Hall (1978); Zaitzev (1989).

Major Taxonomic Works and Lists. Becker (1913a); Bezzi (1924); Coquillett (1910); Evenhuis (1983b); Hall (1976a); Hull (1973); Paramonov (1939b).

In addition to these references, yearly issues of the *Zoological Record* were checked for possible type designations (e.g., an earlier one was found for *Hyperalonia* Rondani) and emendations that might have been missed by other workers.

Classification

THE BOMBYLIIDAE HAVE BEEN SUBJECTED to various classifications since Becker's (1913a) seminal work on the classification of the family. Most authors have dealt with taxa on a limited basis, such as in regional catalogs (Painter & Painter 1965; Bowden 1975c, 1980; Painter, Painter & Hall 1978; Zaitzev 1989; Evenhuis 1989b), or in regional systematic works (Hesse 1938; Zaitzev 1966; Hall & Evenhuis 1980). Bezzi (1924), Paramonov (1939b), Hull (1973), and Hall (1976a) dealt with the classification of the family on a world basis. Of these last 4, Paramonov (1938) was only a reprinting of Bezzi (1924) with the addition of a few genera that had been described in the intervening years. Hall (1976a) continued this listing by adding the genera described since Paramonov (1939b). One unfortunate factor that accompanied these regional classifications is that explanations justifying each author's proposals were often omitted (in the case of regional catalogs this was because of format restrictions). Hull (1973) gave explanations for his classification, but subsequent workers have felt that Hull's classification presented many problems and solved few. Hull created many tribes for some subfamilies and tended to lump other tribes within a large and unwieldy concept of Bombyliinae.

It is sorely evident that a reclassification of the Bombyliidae sensu lato is necessary. This can only be done after detailed analysis of all of the genera comprising the family and comparing them with related families of Diptera. The classification presented here is provisional, and is only intended as a reference for future analyses on this family. I have examined virtually all of the genera listed below and followed the latest published classifications of each subfamily, with only a few changes being made where subsequent research has shown the need for such changes. The use of Homoeophthalmae and Tomophthalmae follows Theodor's concept of these sections based on whether the occipital foramen is single- (Homoeophthalmae) or double-chambered (Tomophthalmae).

I am in the process of revising the classification of the Bombyliidae, with the assistance of Dr. David Greathead, on a subfamily-by-subfamily basis. The Phthiriinae and Usiinae have been completed (Evenhuis 1990a) and the Mythicomyiinae are in preparation. Dr. David K. Yeates is also in the process of reexamining the higher classification of the family and its relationship with other asiloid families. The final results of these studies should provide a solid framework for future systematic studies on the Bombyliidae.

Section Homoeophthalmae
Mythicomyiinae
Acoecus Hull, 1973
Acridophagus Evenhuis, 1983
Cyrtomorpha White, 1916
Robertsella Evenhuis, 1978
Aetheoptilus Hesse, 1967
Anomaloptilus Hesse, 1938
Cephalodromia Becker, 1914
Ceratolaemus Hesse, 1938
Cyrtisiopsis Séguy, 1930
Cyrtoides Engel, 1933
Ecliptica Engel, 1933
Cyrtosia Perris, 1839
Doliopteryx Hesse, 1956
Doliopteryx Hesse, 1938
Empidideicus Becker, 1907
Euanthobates Hesse, 1965
Glabellula Bezzi, 1902
Platygaster Zetterstedt, 1838
Sphaerogaster Zetterstedt, 1842
Glabella Loew, 1873
Pachyneres Greene, 1924
Proglabellula† Hennig, 1966, **n. syn.**
Leylaiya Efflatoun, 1945
Mnemomyia Bowden, 1975
Mythenteles Hall & Evenhuis, 1986, **n. stat.**
Cladella Hull, 1973
Cladella Hall & Evenhuis, 1987, **n. syn.**
Mythicomyia (Mythastrabes) Hall & Evenhuis, 1986
Mythicomyia (Mythicomyia) Coquillett, 1893
Heterohybos Brèthes, 1920
Nexus Hall & Evenhuis, 1987
Onchopelma Hesse, 1938
Palaeoplatypygus† Kovalev, 1985
Paraconsors (Elachymyia) Hall & Evenhuis, 1987
Paraconsors (Paraconsors) Hall & Evenhuis, 1987
Platypygus Loew, 1844
Popsia Costa, 1863
Procyrtosia† Zaitzev, 1986
Proplatypygus† Hennig, 1969
Pseudoglabellula Hesse, 1967
Psiloderoides Hesse, 1967

†Fossil or amber genus.

Gerontinae
Geron (Geron) Meigen, 1820
Amictogeron Hesse, 1938
Geron (Empidigeron) Painter, 1932
Geron (Pseudoammictus) Bigot, 1892
Pseudempis Bezzi, 1921
Geron (Plichtamyia) Evenhuis, 1979
Systropodinae
Dolichomyia Wiedemann, 1830
Melanderella† Cockerell, 1909
Systropus (Diaerops) Enderlein, 1926
Coptodicrus Enderlein, 1926
Systropus (Dimelopelma) Enderlein, 1926
Systropus (Systropus) Wiedemann, 1820
Céphène Latreille, 1825
Cephenes Latreille, 1829
Cephenus Berthold, 1827
Systropus Henning, 1832
Xystropus Agassiz, 1847
Cephenius Enderlein, 1926
Coptopelma Enderlein, 1926
Pioperna Enderlein, 1926, **n. syn.**
Symballa Enderlein, 1926
Systropus (Teinopelmus) Bowden, 1967
Zaclava Hull, 1973
Toxophorinae
Toxophora Meigen, 1803
Eniconevra Macquart, 1840
Heniconevra Agassiz, 1847
Heniconeura Bezzi, 1903
Toxomyia Hull, 1973
Bombyliinae
Acrophthalmydini
Acrophthalmyda Bigot, 1858
Scinax Loew, 1855
Ostentator Jaennicke, 1867
Bombyliini
Bombylius (Bombylius) Linnaeus, 1758
Parisus Walker, 1852
Triplasius Loew, 1855
Bombylius (Parabombylius) Williston, 1907
Bombylius (Zephyrectes) Evenhuis, 1978
Eurycarenus Loew, 1860
Euchariomyia Bigot, 1888
Eucharimyia Bigot, 1889, **n. syn.**
Systoechus Loew, 1855

Conophorini
- **Aldrichia** Coquillett, 1894
- **Conophorina** Becker, 1920
- **Conophorus** Meigen, 1803
  - *Ploas* Latreille, 1804
  - *Tornotes* Gistl, 1848, **n. syn.**
  - *Codionus* Rondani, 1873
  - *Calopelta* Greene, 1921
- **Legnotomyia** Bezzi, 1902
  - *Legnotus* Loew, 1855
  - *Psiatholasius* Becker, 1906
- **Othniomyia** Hesse, 1938
- **Platamomyia** Brèthes, 1926
  - *Platamodes* Loew, 1855
- **Prorachthes** Loew, 1869
  - *Dumontiella* Séguy, 1932
  - *Cheilohadrus* Hesse, 1938
- **Sosiomyia** Bezzi, 1921
- **Sparnopolius** Loew, 1855

Crocidiini
- **Adelogenys** Hesse, 1938
- **Apatomyza** Wiedemann, 1820
  - *Apatomyza* Henning, 1832
- **Crocidium** Loew, 1860
- **Mallophthiria** Edwards, 1930
- **Megaphthiria** Hall, 1976
- **Semiramis** Becker, 1913
- **Timiomyia** Evenhuis, 1978
  - *Tamerlania* Paramonov, 1931

Dischistini
- **Adelidea** Macquart, 1840
  - *Sobarus* Loew, 1855
- **Anastoechus** Osten Sacken, 1877
- **Bombylisoma** Rondani, 1856
  - *Bombylosoma* Marschall, 1873
  - *Chasmoneura* Hesse, 1938
- **Bromoglycis** Hull, 1971
- **Brychosoma** Hull, 1973
- **Cacoplox** Hull, 1970
- **Choristus** Walker, 1852
- **Cryomyia** Hull, 1973
- **Dischistus** Loew, 1855
  - *Acanthogeron* Bezzi, 1925
  - *Lissomerus* Austen, 1937
- **Doliogethes** Hesse, 1938
- **Efflatounia** Bezzi, 1925
- **Eristalopsis** Evenhuis, 1985
  - *Syrphoides* Evenhuis, 1983
- **Euprepina** Hull, 1971
- **Eusurbus** Roberts, 1929
- **Geminaria** Coquillett, 1894
- **Gonarthrus** Bezzi, 1921
- **Hallidia** Hull, 1970
- **Heterostylum** Macquart, 1848
  - *Comastes* Osten Sacken, 1877
  - *Unduloptomyia* Evenhuis, 1978
- **Isocnemus** Bezzi, 1924
- **Karakumia** Paramonov, 1927
- **Laurella** Hull, 1971
- **Lepidochlanus** Hesse, 1938
- **Lordotus** Loew, 1863
- **Mandella** Evenhuis, 1983
- **Meomyia** Evenhuis, 1983
  - *Paramonovella* Hull, 1973
  - *Paramonovella* Zaitzev, 1982
- **Muscatheres** Evenhuis, 1986
- **Nectaropota** Philippi, 1865
- **Neobombylodes** Evenhuis, 1978
  - *Bombylodes* Paramonov, 1940
- **Neodischistus** Painter, 1933
- **Nothoschistus** Bowden, 1985
- **Parachistus** Greathead, 1980
- **Parasystoechus** Hall, 1976
- **Paratoxophora** Engel, 1936
- **Pilosia** Hull, 1973
- **Semistoechus** Hall, 1976
- **Sericusia** Edwards, 1937
- **Sisyromyia** White, 1916
- **Sisyrophanus** Karsch, 1886
- **Staurostichus** Hull, 1973
- **Tovlinius** Zaitzev, 1979
- **Triploechus** Edwards, 1937
- **Zinnomyia** Hesse, 1955

Desmatomyiinae
- **Desmatomyia** Williston, 1895
  - *Acreotrichites*† Cockerell, 1917
- **Inyo** Hall & Evenhuis, 1987

Phthiriinae

Phthiriini
- **Acreophthiria** Evenhuis, 1986
- **Acreotrichus** Macquart, 1850
- **Australiphthiria** Evenhuis, 1986
- **Neacreotrichus** Cockerell, 1917
- **Phthiria** Meigen, 1820
  - *Ptinia* Rafinesque, 1815, **n. syn.**
- **Pygocona** Hull, 1973

Poecilognathini
- **Euryphthiria** Evenhuis, 1986
- **Poecilognathus** Jaennicke, 1867
  - *Agenosia* Hull, 1973
  - *Geronites*† Cockerell, 1915
- **Relictiphthiria** Evenhuis, 1986
- **Tmemophlebia** Evenhuis, 1986
  - *Eccremus* Wiedemann, 1821, **n. syn.**

Cyclorhynchus Macquart, 1840
Cyclorrhynchus Bezzi, 1924
Usiinae
Apolysini
Apolysis Loew, 1860
Pseudogeron Cresson, 1915
Rhabdopselaphus Bigot, 1886
Dagestania Paramonov, 1929
Usiini
Parageron Paramonov, 1929
Usia (Micrusia) Evenhuis, 1990
Usia (Usia) Latreille, 1802
Voluccella Fabricius, 1794
Zerinthia Rafinesque, 1815, **n. syn.**
Ecliminae
Alepidophora† Cockerell, 1909
Cyrtomyia Bigot, 1892
Cyrtophorus Bigot, 1857
Toxophorinarum Verrall, 1909
Eclimus Loew, 1844
Eclimmus Agassiz, 1847
Lepidophora Westwood, 1835
Marmasoma White, 1916
Palintonus François, 1964
Thevenetimyia (Arthroneura) Hull, 1966
Thevenetimyia (Thevenetimyia) Bigot, 1875
Thevenemyia Bigot, 1875
Epibates Osten Sacken, 1877
Tillyardomyia Tonnoir, 1927
Heterotropinae
Heterotropus Loew, 1873
Malthacotricha Becker, 1907
Oligodraninae
Oligodranes Loew, 1844
Incertae Sedis
Pachysystropus† Cockerell, 1909
Palaeogeron† Meunier, 1914
Tithonomyia† Evenhuis, 1984

Section Tomophthalmae
Cythereinae
Cythereini
Callostoma Macquart, 1840
Callistoma Agassiz, 1847
Cytherea (Chalcochiton) Loew, 1844
Cytherea (Cytherea) Fabricius, 1794
Mulio Latreille, 1796
Astropha Rafinesque, 1815, **n. syn.**
Glossista Rondani, 1856
Logcocerius Rondani, 1863
Loncocerius Bezzi, 1903
Lonchocerius Bezzi, 1924
Pantarbes Osten Sacken, 1877
Sericosoma Macquart, 1850
Sericothrix Hall, 1976
Bombylthrix Hall, 1976
Mariobezziini
Callynthrophora Schiner, 1867
Corsomyza Wiedemann, 1820
Corsomyza Henning, 1832
Lasioprosopa Macquart, 1855
Denamyza Hull, 1973
Dasypalpus Macquart, 1840
Dasypalpoides Evenhuis, 1978, **n. syn.**
Gnumyia Bezzi, 1921
Hyperusia Bezzi, 1921
Mariobezzia Becker, 1913
Megapalpus Macquart, 1834
Neosardus Roberts, 1929
Paracorsomyza† Hennig, 1966
Pusilla Paramonov, 1954
Zyxmyia Bowden, 1960
Tribus inquirendus
Gyrocraspedum Becker, 1913
Praecytherea† Théobald, 1937
Xenoprosopinae
Xenoprosopa Hesse, 1956
Oniromyiinae
Oniromyia Bezzi, 1921
Xethomyza Hesse, 1938
Cylleniinae
Cylleniini
Amictites† Hennig, 1966
Amictus Wiedemann, 1817
Thlipsomyza Wiedemann, 1820
Thlipsomyza Henning, 1832
Thlipsogaster Rondani, 1863
Truquia Rondani, 1863
Thlipsigaster Marschall, 1873
Subbombylius Kapoor, 1979, **n. syn.**
Cyllenia Latreille, 1802
Glaesamictus† Hennig, 1966
Palaeoamictus† Meunier, 1916
Sinaia Becker, 1916
Sphenoidoptera Williston, 1901
Enicini
Enica Macquart, 1834
Henica Agassiz, 1847

Lagochilus Loew, 1860
Alonipola Rondani, 1863
Nomalonia Rondani, 1863
Peringueyimyiini
Peringueyimyia Bigot, 1886
Tribus inquirendus
***Lithocosmus*†** Cockerell, 1909
Protophthiria† Cockerell, 1914
Tomomyzinae
Amphicosmus (Amphicosmus) Coquillett, 1891
Amphicosmus (Glycophorba) Hull, 1973
Docidomyia White, 1916
Megacosmus† Cockerell, 1909
Metacosmus Coquillett, 1891
Myonema Roberts, 1929
Pantostomus Bezzi, 1921
Paracosmus (Actherosia) Hull, 1973
Paracosmus (Paracosmus) Osten Sacken, 1877
Allocotus Loew, 1872
Protolomatia† Cockerell, 1914, **n. syn.**
Tomomyza Wiedemann, 1820
Tomomyza Henning, 1832
Antoniinae
Antonia Loew, 1856
Dimorphaphus Walker, 1871
Dimorphophora Walker, 1871
Antoniaustralia Becker, 1913
Lomatiinae
Comptosiini
Aleucosia Edwards, 1934
Chelina Hull, 1973
Paramonovomyia Evenhuis, 1980
Comptosia Macquart, 1840
Neuria Newman, 1841
Nevria Agassiz, 1847, **n. syn.**
Alyosia Rondani, 1863
Anthocolon Hull, 1973
Epidosia Hull, 1973
Opsonia Hull, 1973
Paradosia Hull, 1973
Doddosia Edwards, 1934
Macrocondyla (Lyophlaeba) Rondani, 1863
Tritoneura Schiner, 1867
Liophlaeba Bezzi, 1924
Compsella Hull, 1973
Macrocondyla (Macrocondyla) Rondani, 1863
Oncodosia Edwards, 1934
Ulosometa Hull, 1973
Ylasoia Speiser, 1920
Lomatiini
Anisotamia Macquart, 1840
Ancentralis Bezzi, 1925
Brachydemia Hull, 1973
Bryodemina Hull, 1973
Bryodema Hull, 1966
Canariellum Strand, 1928
Canaria Becker, 1913
Edmundiella Becker, 1915
Lomatia Meigen, 1822
Stygia Meigen, 1820
Stygide Latreille, 1825
Stygidium Berthold, 1827, **n. syn.**
Stygides Latreille, 1829, **n. syn.**
Nettadion Gistl, 1848, **n. syn.**
Ogcodocera Macquart, 1840
Oncodeocera Agassiz, 1847
Oncodocera Verrall, 1882
Incertae sedis
Alomatia† Cockerell, 1914
Anthracinae
Anthracini
Anthrax Scopoli, 1763
Chalcamoeba Sack, 1909
Brachyanax Evenhuis, 1981
Coniomastix Enderlein, 1934
Dicranoclista Bezzi, 1924
Coquillettia Williston, 1896
Satyramoeba Sack, 1909
Meganthrax Evenhuis, 1979
Spogostylum Macquart, 1840
Spongostylum Agassiz, 1847
Agryromoeba Schiner, 1860
Argyramoeba Loew, 1869, **n. syn.**
Anthracamoeba Sack, 1909
Chrysamoeba Sack, 1909
Leucamoeba Sack, 1909
Molybdamoeba Sack, 1909
Psamatamoeba Sack, 1909
Aureomoeba Evenhuis, 1978
Turkmeniella Paramonov, 1939
Walkeromyia Paramonov, 1934
Xenox Evenhuis, 1984
Stymphalina Hull, 1973
Aphoebantini
Aphoebantus Loew, 1872
Triodites Osten Sacken, 1877
Triodina Hull, 1973
Meganedys Hull, 1973, **n. syn.**

Epacmus Osten Sacken, 1886
Leptochilus Loew, 1872
Eucessia Coquillett, 1886
Exepacmus Coquillett, 1894
Cononedys Hermann, 1907
Conogaster Hermann, 1907
Pteraulax Bezzi, 1921
Pteraulacodes Hesse, 1956
Exoprosopini
Exoprosopa Macquart, 1840
Litorhynchus Macquart, 1840
Trinaria Mulsant, 1852
Argyrospila Rondani, 1856
Defilippia Lioy, 1864
Litorrhynchus Verrall, 1882, **n. syn.**
Exoptata Coquillett, 1887
Corycetta Hull, 1973
Litomyza Hull, 1973
Atrichochira Hesse, 1956
Collosoptera Hull, 1973
Diatropomma Bowden, 1962
Heteralonia (Heteralonia) Rondani, 1863
Zygodipla Bezzi, 1924
Heteralonia (Homolonia) Bowden, 1975
Mima Meigen, 1820
Heteralonia (Isotamia) Bezzi, 1912
Francoisia Hesse, 1952
Ogilviella Paramonov, 1954
Stenolonia Bowden, 1975
Heteralonia (Acrodisca) Bezzi, 1921
Cladodisca Bezzi, 1922
Heteralonia (Mesoclis) Bezzi, 1921
Heteralonia (Metapenta) Bezzi, 1921
Ligyra (Ligyra) Newman, 1841
Velocia Coquillett, 1886
Paranthrax Paramonov, 1931
Paranthracina Paramonov, 1933
Ligyra (Hyperalonia) Rondani, 1863
Litorhina Bowden, 1975
Exosoma Séguy, 1938
Micomitra Bowden, 1964
Pterobates Bezzi, 1921
Petrorossiini
Desmatoneura Williston, 1895
Chionamoeba Sack, 1909
Chiasmella Bezzi, 1924
Petrorossia Bezzi, 1908
Pipunculopsis Bezzi, 1925
Prothaplocnemis Bezzi, 1925
Xeramoeba Hesse, 1956
Plesiocerini
Conomyza Hesse, 1956
Coryprosopa Hesse, 1956
Epacmoides Hesse, 1956
Plesiocera Macquart, 1840
Calledax Hull, 1973
Prorostoma Hesse, 1956
Stomylomyia Bigot, 1887
Villini
Astrophanes Osten Sacken, 1877
Caecanthrax Greathead, 1981
Caecanthrax Greathead, 1980
Chrysanthrax Osten Sacken, 1886
Cyananthrax Painter, 1959
Deusopora Hull, 1971
Diochanthrax Hall, 1975
Dipalta Osten Sacken, 1877
Diplocampta Schiner, 1867
Exechohypopion Evenhuis, 1991
Exhyalanthrax Becker, 1916
Oriellus Hull, 1973
Tauropsis Hull, 1973, **n. syn.**
Hemipenthes Loew, 1869
Isopenthes Osten Sacken, 1886
Laminanthrax Greathead, 1967
Lepidanthrax Osten Sacken, 1886
Podolepida Hull, 1973
Mancia Coquillett, 1886
Neodiplocampta (Agitonia) Hull, 1966
Neodiplocampta (Neodiplocampta) Curran, 1934
Pachyanthrax François, 1964
Paradiplocampta Hall, 1975
Paranthrax Bigot, 1876
Paravilla Painter, 1933
Poecilanthrax Osten Sacken, 1886
Pseudopenthes Roberts, 1928
Rhynchanthrax Painter, 1933
Stonyx Osten Sacken, 1886
Synthesia Bezzi, 1921
Thyridanthrax Osten Sacken, 1886
Veribubo Evenhuis, 1978
Tumulus Hull, 1973
Verrallites† Cockerell, 1913
Villa Lioy, 1864
Hyalanthrax Osten Sacken, 1886
Aspiloptera Künckel, 1905
Protepacmus† Cockerell, 1916, **n. syn.**

Villoestrini

Marleyimyia Hesse, 1956
Oestranthrax Bezzi, 1921
Oestrimyza Hull, 1973
Villoestrus Paramonov, 1931

?Family Bombyliidae

Proratinae

Prorates Melander, 1906
Alloxytropus Bezzi, 1925
Apystomyia Melander, 1950
Caenotus Cole, 1923
Caenotoides Hall, 1972

Incertae Sedis

Anthracida† Germar, 1839
Apsilocephala Kröber, 1914
Clesthentia White, 1914
Hilarimorpha Schiner, 1860

Genus-group Names of Bombyliidae

Acanthogeron Bezzi, 1925:164. Type species: *Bombylius senex* Wiedemann *in* Meigen, 1820, by original designation. Synonymized with *Dischistus* Loew, 1855, by Bowden (1973:140).

Remarks. Theodor (1983:111) resurrected *Acanthogeron* as a subgenus of *Dischistus,* but I feel this action is premature and follow Bowden's (1973) synonymization of it under *Dischistus.*

Aceotrichus: Macquart, 1850:470(166). Incorrect original spelling. See *Acreotrichus* Macquart.

Acoecus Hull, 1973:265, 276 (as subgenus of *Euanthobates* Hesse). Type species: *Euanthobates mellivorus* Hesse, 1967, by original designation. Raised to generic status by Bowden (1980:382). **Mythicomyiinae.**

Acreophthiria Evenhuis, 1986:6. Type species: *Acreotrichus americanus* Coquillett, 1895, by original designation. **Phthiriinae: Phthiriini.**

Acreotichus: Macquart, 1850:310(6). Incorrect original spelling. See *Acreotrichus* Macquart.

Acreotrichites† Cockerell, 1917:376. Type species: *Acreotrichites scopulicornis* Cockerell, 1917, by original designation. Synonymized with *Desmatomyia* Williston, 1895, by Evenhuis (1990a:65). Fossil shale genus.

Acreotrichus Macquart, 1850:425(121). Type species: *Acreotrichus gibbicornis* Macquart, 1850 [= *Bombylius antecedens* Walker, 1849], by original designation. **Phthiriinae: Phthiriini.**

Remarks. Macquart used 3 different spellings for this genus in his original publication (*Acreotichus* on p. 310; *Aceotrichus* on p. 470; *Acreotrichus* on p. 425). Evenhuis

(1986:7) decided that *Aceotrichus* was an incorrect original spelling of *Acreotrichus* Macquart, 1850. I consider here *Acreotichus* to be an incorrect original spelling of *Acreotrichus* Macquart, 1850.

Acridophagus Evenhuis, 1983c:318 (new replacement name for *Robertsella* Evenhuis). Type species: *Cyrtomorpha paganica* White, 1916, automatic. **Mythicomyiinae.**

Acrodisca Bezzi, 1921a:138, 139 [1922c:85; 1923:332; 1924:234] (as subgenus of *Exoprosopa* Macquart). Type species: *Exoprosopa angulata* Loew, 1860, by subsequent designation of Bezzi (1924:234). Treated as a subgenus of *Heteralonia* Rondani, 1863, by Bowden (1975b:318). **Anthracinae: Exoprosopini.**

Remarks. Bezzi described this genus-group name as new 4 times (1921a:138; 1922c:84–85; 1923:332; and 1924:234). The original description in 1921 included 4 species (*fimbriatella*, n. sp., *offuscata*, n. sp., *personata*, n. sp., and *angulata* Loew) but did not have a type species designation. The 2nd mention (1922c:84–85) included 2 species (*angulata* Loew and *spoliata*, n. sp.) without designating a type. The 3rd description (1923:332) included only 1 species (*suffusipennis*, n. sp.) (not one of the originally included species in 1921). Bezzi (1924:234) included 12 species and designated *angulata* as the type species, which is the 1st valid subsequent type-designation of an originally included species.

Acrophthalmyda Bigot, 1858:573, 583. Type species *Cyllenia elegantula* Bigot, 1857, by monotypy. **Bombyliinae: Acrophthalmydini.**

Remarks. Scudder (1882a:5), based on a list of Diptera genus-group names given to him by Bigot, gives this genus as "Acropthalmyia" with the corresponding derivation in Greek. However, this does not satisfy the I.C.Z.N. criteria for it to be a true emendation. See Verrall (1889:265) for discussion of Bigot's orthography and a portion of this list supplied to Scudder.

Actherosia Hull, 1973:298 (as subgenus of *Paracosmus* Osten Sacken). Type species *Paracosmus rubicundus* Melander, 1950, by original designation. **Tomomyzinae.**

Adelidea Macquart, 1840:84 [1841:362]. Type species: *Adelidea fuscipennis* Macquart, 1840 [= *Bombylius anomalus* Wiedemann, 1821], by original designation. **Bombyliinae: Dischistini.**

Adelogenys Hesse, 1938:811. Type species: *Adelogenys culicoides* Hesse, 1938, by original designation. **Bombyliinae: Crocidiini.**

Aetheoptilus Hesse, 1967:112 (as subgenus of *Empidideicus* Becker). Type species: *Empidideicus zuluensis* Hesse, 1967, by original designation. Raised to generic status by Bowden (1980:382). **Mythicomyiinae.**

Remarks. This genus was originally distinguished on the basis of the presence of a marginal cell in the wing. Examination of the type specimen shows this feature not to be present, hence its status as a genus or subgenus may be in doubt pending further investigation.

Agenosia Hull, 1973:195 (as subgenus of *Phthiria* Meigen). Type species: *Phthiria vittata* Hull, 1973 [= *Phthiria loewi* Painter, 1962], by original designation. Synonymized with *Poecilognathus* Jaennicke, 1867, by Evenhuis (1986:14).

Agitonia Hull, 1966:225 (as subgenus of *Neodiplocampta* Curran). Type species: *Neodiplocampta sepia* Hull, 1966, by monotypy. **Anthracinae: Villini.**

Aldrichia Coquillett, 1894:93. Type species: *Aldrichia ehrmanii* Coquillett, 1894, by monotypy. **Bombyliinae: Conophorini.**

Alepidophora† Cockerell, 1909b:54. Type species: *Alepidophora pealei* Cockerell, 1909, by monotypy. Fossil shale genus. **Ecliminae.**

Aleucosia Edwards, 1934:95. Type species: *Neuria corculum* Newman, 1841, by original designation. **Lomatiinae: Comptosiini.**

Allocotus Loew, 1872a:82 [1872b:258]. Type species: *Allocotus edwardsii* Loew, 1872, by monotypy. Junior homonym of *Allocotus* Fisher von Waldheim, 1838. Objective synonym of *Paracosmus* Osten Sacken, 1877.

Alloxytropus Bezzi, 1925:184, 186. Type species: *Alloxytropus anomalus* Bezzi, 1925, by original designation. **Proratinae.**

Remarks. Engel (1937:516) first recognized the similarity between *Alloxytropus* and *Prorates* Melander, 1906, and said it was highly probable that the 2 genera were synonymous. Melander (1950:145), Hall (1972:38), and Bowden (1980:430) formally synonymized *Alloxytropus* with *Prorates,* yet Zaitzev (1989:58) preferred to retain *Alloxytropus* as a distinct genus. Until recently, *Prorates* and *Alloxytropus* were known only from small populations in the southwestern United States, Kazakhstan, Mongolia, the Sudan, and Egypt. I have seen more undescribed material of Proratinae from Argentina and Australia and prefer to follow Zaitzev (1989) in maintaining *Alloxytropus* and *Prorates* as separate genera until this subfamily is better known worldwide. See also remarks under *Caenotoides* Hall.

Alomatia† Cockerell, 1914b:721. Type species: *Alomatia fusca* Cockerell, 1914, by monotypy. Fossil shale genus. **Lomatiinae: *Tribus inquirendus.***

Remarks. I examined the type of *Alomatia fusca* in the National Museum of Natural History, Washington, D.C. (USNM). All that remains are fragments of the surrounding matrix. The type itself could not be found among the debris. Until the type or other conspecific specimens are found, the genus must remain *tribus inquirendus*.

Alonipola Rondani, 1863:71 [1864:71]. Type species *Cyllenia pluricellata* Macquart, 1855 [= *Anthrax longirostris* Wiedemann, 1819], by original designation. Synonymized with *Enica* Macquart, 1834, by Kertész (1909:69).

Alyosia Rondani, 1863:54 [1864:54]. Type species: *Comptosia maculipennis* Macquart, 1846 [= *Neuria ocellata* Newman, 1841], by subsequent designation of Becker (1913a:465). Synonymized under *Comptosia* Macquart, 1840, by Yeates (in litt.).

Amictites† Hennig, 1966:9. Type species: *Amictites regiomontana* Hennig, 1966, by monotypy. Baltic amber genus. **Cylleniinae: Cylleniini.**

Remarks. This is the only fossil bombyliid genus where a pupal exuvium is known.

Amictogeron Hesse, 1938:918 (as genus). Type species: *Amictogeron meromelanus* Hesse, 1938, by original designation. Synonymized with the subgenus *Geron (Geron)* Meigen, 1820, by Bowden (1974:91).

Amictus Wiedemann, 1817:58. Type species: *Bombylius oblongus* Fabricius, 1805, by monotypy. **Cylleniinae: Cylleniini.**

Amphicosmus Coquillett, 1891:219. Type species: *Amphicosmus elegans* Coquillett, 1891, by monotypy. **Tomomyzinae.**

Remarks. According to Hull (1973:295), 2 subgenera are recognized: subg. **Amphicosmus** s. str. and subg. **Glycophorba** Hull, 1973.

Anastoechus Osten Sacken, 1877:251. Type species: *Anastoechus barbatus* Osten Sacken, 1877, by monotypy. **Bombyliinae: Dischistini.**

Remarks. Previous classifications have placed this genus in the tribe Bombyliini (e.g., Bowden 1985). The absence of hairs on the metapleuron (= laterotergite) precludes *Anastoechus* from being a member of the Bombyliini as currently defined. It is here placed in the Dischistini until further analysis is completed to ascertain its position in the Bombyliinae.

Ancentralis: Bezzi, 1925:199. Type species: *Lomatia pictipennis* Wiedemann, 1828, by monotypy. Unavailable name. Proposed in specific synonymy with *Lomatia pictipennis* Wiedemann, 1828; not made available before 1961. Synonym of *Anisotamia* Macquart, 1840.

Remarks. Though no evidence exists as such in the original publication, this name was actually a typographical error for "An. centralis" referring to *Anisotamia centralis* Macquart, 1840.

Anisotamia Macquart, 1840:81 [1841:359]. Type species: *Anisotamia ruficornis* Macquart, 1840, by subsequent designation of Coquillett (1910:507). **Lomatiinae: Lomatiini.**

Anomaloptilus Hesse, 1938:983 (as subgenus of *Empidideicus* Becker). Type species: *Empidideicus celluliferus* Hesse, 1938, by monotypy. Synonymized with *Empidideicus* Becker, 1907, by Bowden (1980:383).

Anthax: Verrall *in* Scudder, 1882a:26. See *Anthrax* Scopoli.

Anthocolon Hull, 1973:307 (as subgenus of *Comptosia* Macquart). Type species: *Comptosia casimira* Hull, 1973, by original designation. Raised to generic status by Evenhuis (1989b:374). Synonymized under *Comptosia* Macquart, 1840, by Yeates (in litt.).

Anthracida† Germar, 1849:64. Type species: *Anthracida xylotona* Germar, 1849, by monotypy. Fossil shale genus. ***Incertae sedis* (within Diptera).**

Remarks. Neave (1939a:228) attributed *Anthracida* to Coleoptera, yet Scudder (1882a:25) treated it within the Diptera and Handlirsch (1908:1011), Schulze, Kükenthal, and Heider (1926:210), Paramonov (1939b:27, 42), and Hull (1973:59) all treated it within the Bombyliidae. The original description mentions it as allied to Anthracinae, but the wing venation does not appear to represent a bombyliid. Correct familial placement will have to await examination of the type fossil.

Anthracamoeba Sack, 1909:515. Type species: *Anthracamoeba obscurum* Sack, 1909, by original designation. Synonymized with *Argyromoeba* Schiner, 1860, by Becker (1913a:448). Synonymized with *Spogostylum* Macquart, 1840, by Engel (1936a:438).

Anthrax Scopoli, 1763:358. Type species: *Musca morio* Linnaeus, 1758 [misidentification, = *Musca anthrax* Schrank, 1781], by monotypy. **Anthracinae: Anthracini.**

Remarks. Because both *Hemipenthes* Loew, 1869, and *Anthrax* are based upon the same type species, one or the other name would have to be sunk to resolve the nomenclatural conflict. An application to the I.C.Z.N. is in preparation to fix the type species by plenary power and resolve the situation. The names *Anthax* and *Antrax,* listed by Scudder (1882a), are subsequent misspellings of *Anthrax* and are not considered emendations.

Antonia Loew, 1856:30. Type species: *Antonia suavissima* Loew, 1856, by monotypy. **Antoniinae.**

Antoniaustralia Becker, 1913a:458. Type species: *Antoniaustralia hermanni* Becker, 1913, by original designation. **Antoniinae.**

Antrax: Verrall *in* Scudder, 1882a:25. See *Anthrax* Scopoli.

Apatomyza Henning, 1832:326. Type species: *Apatomyza punctipennis* Henning, 1832 [preoccupied, = *Apatomyza punctipennis* Wiedemann, 1820], by monotypy. Junior homonym of *Apatomyza* Wiedemann, 1820. Synonymized with *Apatomyza* Wiedemann, 1820, implicitly by Hagen (1862:356) and explicitly by Evenhuis (1990a:60).

Remarks. Hagen (1862:356), by stating that Henning's (1832) was a plagiarized copy of Wiedemann (1820), thus implicitly synonymizes all the taxa therein treated as new with those in Wiedemann (1820).

Apatomyza Wiedemann, 1820:11. Type species: *Apatomyza punctipennis* Wiedemann, 1820, by original designation (on plate). **Bombyliinae: Crocidiini.**

Aphoebantes: Bigot *in* Scudder, 1882a:27. See *Aphoebantus* Loew.

Aphoebantus Loew, 1872a:77 [1872b:252]. Type species: *Aphoebantus cervinus* Loew, 1872, by monotypy. **Anthracinae: Aphoebantini.**

Remarks. Scudder (1882a:27), based on a list of Diptera given to him by Bigot, gives the spelling as *Aphoebantes.* This, however, does not satisfy the criteria for a true emendation.

Apolysis Loew, 1860a:86 [1860b:269]. Type species: *Apolysis humilis* Loew, 1860, by monotypy. **Usiinae: Apolysini.**

Apsilocephala Kröber, 1914:36. Type species: *Apsilocephala longistylus* Kröber, 1914, by monotypy. ***Incertae sedis* (within non-cyclorrhaphous Diptera).**

Remarks. Irwin (1976:921) stated that this genus was an anomalous therevid and allied it to *Prorates* Melander. Irwin and Lyneborg (1981:517) placed it close to the genera allied to *Caenotus* Cole. I found in preliminary examination of the type specimen that it has antennal characteristics (2-segmented arista) more typical of Empidoidea and 5 posterior cells in the wing. At present, it is still an enigmatic genus. More detailed study is necessary to ascertain its correct family placement.

Apystomyia Melander, 1950:146. Type species: *Apystomyia elinguis* Melander, 1950, by original designation. **?Bombyliidae: Proratinae.**

Remarks. Currently, this genus is an enigma in the family Bombyliidae. It exhibits similarities to Empidoidea as well as genera allied to *Prorates* Melander. Until a better definition of the limits of the Bombyliidae, sensu stricto, can be formulated, *Apystomyia* is placed in Proratinae. See also remarks under *Caenotoides* Hall.

Argyramoeba Loew, 1869b:228 (unjustified emendation of *Argyromoeba* Schiner). Type species: *Anthrax tripunctata* Wiedemann, 1821, automatic. Synonym of *Spogostylum* Macquart, 1840, **new synonymy.**

Remarks. Loew originally satisfied the I.C.Z.N. criteria for an emendation in 1869 but formally emends the genus again [1872a:114; 1872b:290]. *Argyramoeba* of Verrall *in* Scudder (1882a:32), Bezzi (1903:170), Kertész (1909:59), and Bezzi (1924:4) are later emendations of *Argyromoeba.* Some nomenclators and catalogs (e.g., Neave 1939a:292; Hull 1973:436) listed *Argyramoeba* as an emendation or variant spelling attributed to Coucke and Coucke (1894). There is nothing present in that paper to satisfy the criteria of an emendation according to the I.C.Z.N. Code and the spelling is actually *"Argyromaeba"* as correctly cited by Kertész (1909:59). See also remarks under *Argyromoeba* Schiner.

Argyromaeba: Coucke and Coucke, 1894:286. See *Argyramoeba* Loew.

Argyromoeba Schiner, 1860a:51. Type species: *Anthrax tripunctata* Wiedemann, 1821, by subsequent designation of Coquillett (1910:510). Synonymized with *Spogostylum* Macquart, 1840, by Williston (1908:214).

Remarks. Considerable confusion has existed regarding the status of *Argyromoeba* and related genera of Anthracini. The reader is referred to Evenhuis (1985a) for a concise history and clarification of the genera of Anthracini and their respective type species.

Argyrospyla: Rondani, 1856:162. Incorrect original spelling. See *Argyrospila* Rondani.

Argyrospila Rondani, 1856:162, 202. Type species: *Anthrax jacchus* Fabricius, 1805 (as "*jaccus*"), by original designation. Junior homonym of *Argyrospila* Herrich-Schaeffer, 1847. Synonymized with *Exoprosopa* Macquart, 1840, by Loew (1860b:248).

Remarks. Rondani used 2 different spellings of this name in the original publication (*Argyrospyla* on p. 162; *Argyrospila* on p. 202). Verrall's action in Scudder (1882a:32) is considered here to be the 1st revision of the 2 spellings, in which he treated *Argyrospila* as the correct spelling. Bezzi (1924:4) is a later revision of the 2 spellings with *Argyrospila* chosen as the correct spelling.

Arthroneura Hull, 1966:226 (as subgenus of *Eclimus* Loew). Type species: *Eclimus tridentatus* Hull, 1966, by original designation. Treated as a subgenus of *Thevenemyia* Bigot, 1875, by Hall (1969:12–13). Treated as a subgenus of *Thevenetimyia* Bigot, 1875, herein. **Ecliminae.**

Aspiloptera Künckel d'Herculais, 1905:145. Type species: *Anthrax flavus* Meigen, 1820, by monotypy. Synonymized with *Villa* Lioy, 1864, by Paramonov (1939b:31, 36).

Astropha Rafinesque, 1815:130 (unnecessary new replacement name for *Cytherea* Fabricius and *Mulio* Latreille). Type species: *Cytherea obscura* Fabricius, 1794, automatic. Synonym of *Cytherea* Fabricius, 1794, **new synonymy.**

Astrophanes Osten Sacken, 1886a:76 (November) [1886b:106 (December)]. Type species: *Astrophanes adonis* Osten Sacken, 1886, by subsequent monotypy. **Anthracinae: Villini.**

Remarks. Traditional listings of this genus date from its appearance in Osten Sacken (1886b:106). The fact that the *Biologia Centrali-Americana* series appeared in fascicles seems to have eluded previous catalogers of Bombyliidae and results in changes of the previously listed mode of type-species fixations for many genera appearing throughout the series. The name *Astrophanes* is validated first on page 76 (November 1886 fascicle) in the key to genera. It was listed without species included, thus the type species *Astrophanes adonis* Osten Sacken, 1886, is by subsequent monotypy on page 106.

Atrichochira Hesse, 1956b:911. Type species: *Exoprosopa pediformis* Bezzi, 1921, by original designation. **Anthracinae: Exoprosopini.**

Aureomoeba Evenhuis, 1978b:247 (new replacement name for *Chrysamoeba* Sack). Type species: *Chrysamoeba vulpina* Sack, 1909, automatic. Synonymized with *Spogostylum* Macquart, 1840, by Zaitzev (1989:129).

Australiphthiria Evenhuis, 1986:8. Type species: *Phthiria hilaris* Walker, 1852, by original designation. **Phthiriinae: Phthiriini.**

Bombyliosoma: Verrall *in* Scudder, 1882a:47. See *Bombylisoma* Rondani.

Bombylisoma Rondani, 1856:164. Type species: *Bombylius sulphureus* Fabricius, 1805 [= *Bombylius minimus* Scopoli, 1772], by original designation. **Bombyliinae: Dischistini.**

Remarks. Acceptance of a type species designation for *Dischistus* by Rondani (1861:8) of *Bombylius sulphureus* Fabricius would result in *Bombylisoma* becoming an objective junior synonym of *Dischistus* Loew, 1855 (see Evenhuis 1991:164 for details). An application to the I.C.Z.N. to request suppression of all type designations for *Dischistus* previous to Becker (1913a) has been submitted. Scudder (1882a:47), based on a list supplied to him by Verrall, listed the spelling *Bombyliosoma*. This is merely a mention of a subsequent misspelling and is not considered an emendation.

Bombylius Linnaeus, 1758:606. Type species: *Bombylius major* Linnaeus, 1758, by subsequent designation of Latreille (1810:443). **Bombyliinae: Bombyliini.**

Remarks. Latreille's (1810) "Table des genres avec l'indication de l'espèce qui leur sert de type" was considered by Opinion 11 (I.C.Z.N. 1910a) as satisfying the criteria for subsequent designation of type species. According to recent treatment of the genus (Hall & Evenhuis, in press), there are 3 subgenera recognized: subg. **Bombylius** s. str., subg. **Parabombylius** Williston, 1907, and subg. **Zephyrectes** Evenhuis, 1978.

Bombylodes Paramonov, 1940:65, 339. Type species: *Dischistus multisetosus* Loew, 1857, by original designation. Junior homonym of *Bombylodes* Fauvel, 1904. Synonymized with *Karakumia* Paramonov, 1927, by Bowden (1985:99). Treated as a junior objective synonym of *Neobombylodes* Evenhuis, 1978, by Zaitzev (1989:101).

Bombylosoma Marschall, 1873:373 (unjustified emendation of *Bombylisoma* Rondani).Type species: *Bombylius sulphureus* Fabricius, 1805 [= *Bombylius minimus* Scopoli, 1772], automatic. Synonymized with *Bombylisoma* Rondani, 1856, by Hull (1973:107).

Bombylthrix: Hall, 1976a:10. Incorrect original spelling. See *Sericothrix* Hall, 1976.

Brachyanax Evenhuis, 1981:190. Type species: *Brachyanax thelestrephones* Evenhuis, 1981, by original designation. **Anthracinae: Anthracini.**

Brachydemia Hull, 1973:306, 314 (as subgenus of *Bryodemina* Hull). Type species: *Bryodemina latisoma* Hull, 1973, by original designation. Raised to generic status by Painter, Painter and Hall (1978:25). **Lomatiinae: Lomatiini.**

Remarks. Two different spellings of this genus-group name are found in the original publication (*Brachydemia* on p. 306, 314; *Brachydemina* on p. 316). I consider *Brachydemina* to be an incorrect original spelling of *Brachydemia.*

Brachydemina: Hull, 1973:316. Incorrect original spelling. See *Brachydemia* Hull.

Bromoglycis Hull, 1971a:5. Type species: *Bromoglycis robustus* Hull, 1971, by monotypy. **Bombyliinae: Dischistini.**

Brychosoma Hull, 1973:102. Type species: *Brychosoma pulchra* Hull, 1973 [= *Bombylius pictipennis* Macquart, 1850], by original designation. **Bombyliinae: Dischistini.**

Bryodema Hull, 1966:227. Type species: *Anthrax validus* Wiedemann, 1821, by original designation. Junior homonym of *Bryodema* Fieber, 1853. Objective synonym of *Bryodemina* Hull, 1973.

Bryodemina Hull, 1973:314 (new replacement name for *Bryodema* Hull). Type species: *Anthrax validus* Wiedemann, 1821, automatic. **Lomatiinae: Lomatiini.**

Cacoplox Hull, 1970:163. Type species: *Cacoplox griseata* Hull, 1970, by original designation. **Bombyliinae: Dischistini.**

Remarks. Cacoplox was reduced to subgeneric status under *Dischistus* by Painter, Painter and Hall (1978:8) and returned to generic status by Bowden (1985:101, 103). Bowden's treatment of it as a genus is followed here.

Caecanthrax: Greathead, 1980:296, 322–23. Type species: *Anthrax inauratus* Klug, 1832, by present designation. Unavailable name. Genus-group name proposed after 1930 without type species designation. Objective synonym of *Caecanthrax* Greathead, 1981.

Caecanthrax Greathead, 1981:324. Type species: *Anthrax inauratus* Klug, 1832, by original designation. **Anthracinae: Villini.**

Caenotoides Hall, 1972:46. Type species: *Caenotoides californica* Hall, 1972, by original designation. **Proratinae.**

Remarks. This genus, along with *Caenotus* Cole, *Apystomyia* Melander, *Prorates* Melander, and *Alloxytropus* Bezzi, form an uncomfortable group of genera that probably warrant placement in another family (or families). More taxonomic and phylogenetic work needs to be done on these genera to ascertain their correct status. However, until then, I prefer to retain them all within a broad concept of Proratinae sensu Theodor (1983).

Caenotus Cole, 1923:14. Type species: *Caenotus inornatus* Cole, 1923, by original designation. **Proratinae.**

Remarks. See remarks under *Caenotoides* Hall.

Calledax Hull, 1973:338 (as subgenus of *Plesiocera* Macquart). Type species: *Plesiocera psammophila* Hesse, 1956, by original designation. Synonymized with *Plesiocera* Macquart, 1840, by Bowden (1980:429).

Callistoma Agassiz, 1847:60 (unjustified emendation of *Callostoma* Macquart). Type species: *Callostoma fascipennis* Macquart, 1840, automatic. Objective synonym of *Callostoma* Macquart, 1840.

Callostoma Macquart, 1840:77 [1841:355]. Type species: *Callostoma fascipennis* Macquart, 1840, by original designation. **Cythereinae: Cythereini.**

Callynthrophora Schiner, 1867:313 [1868:139]. Type species: *Callynthrophora capensis* Schiner, 1867, by original designation. **Cythereinae: Mariobezziini.**

Calopelta Greene, 1921:23. Type species: *Calopelta fallax* Greene, 1921, by monotypy. Synonymized with *Conophorus* Meigen, 1803, by Engel (1932a:28).

Camptosia: Macquart, 1840:pl. 14, fig. 1. Incorrect original spelling. See *Comptosia* Macquart.

Canaria Becker, 1913a:462. Type species: *Lomatia brunnipennis* Macquart, 1839, by original designation. Junior homonym of *Canaria* Partington, 1835. Objective synonym of *Canariellum* Strand, 1928.

Canariellum Strand, 1928:48 (new replacement name for *Canaria* Becker). Type species: *Lomatia brunnipennis* Macquart, 1839, automatic. **Lomatiinae: Lomatiini.**

Cephalodromia Becker, 1914:121. Type species: *Cephalodromia curvata* Becker, 1914, by monotypy. **Mythicomyiinae.**

Céphène Latreille, 1825:496 (unjustified new replacement name for *Systropus* Wiedemann). Unavailable name. Vernacular name without status in nomenclature.

Remarks. Latreille (1825) thinking that *Systropus* Wiedemann was preoccupied by *Systropha* Illiger, 1806, proposed a new replacement name for it, Céphène. Because Latreille proposed the name in vernacular form and did not Latinize it elsewhere in his 1825 work, it is an unavailable name.

Cephenes: Latreille, 1829:505. Type species: *Systropus macilentus* Wiedemann, 1820, by present designation of 1st species included herein. Unavailable name. Originally proposed in synonymy with *Systropus* Wiedemann, 1820; not made available before 1961. Objective synonym of *Systropus* Wiedemann, 1820.

Remarks. Cephenes Latreille (1829:505) is listed by Neave (1939a:630) as an incorrect subsequent spelling of *Cephenus* Berthold, 1827. However, a check of Latreille's (1829) work shows that it was originally proposed in synonymy with *Systropus* Wiedemann and is not a subsequent misspelling of *Cephenus* Berthold.

Cephenius Enderlein, 1926:70. Type species: *Systropus studyi* Enderlein, 1926, by original designation. Synonymized with the subgenus *Systropus (Systropus)* Wiedemann, 1820, by Bowden (1967:144).

Cephenus Berthold, 1827:506 (unjustified new replacement name for *Systropus* Wiedemann [as "*Systrophus*"]; erroneously deemed preoccupied by *Systropha* Illiger, 1806). Type species: *Systropus macilentus* Wiedemann, 1820, automatic. Objective synonym of *Systropus* Wiedemann, 1820.

Remarks. Berthold's (1827) *Cephenus* is the Latinization of Latreille's (1825) Céphène; the latter is vernacular, thus unavailable. Authorship of the taxa Latinized in Berthold's (1827) translation of Latreille (1825) is attributed to Berthold when they are additional to material that was originally presented in Latreille (1825). Schulze et al. (1927:593) listed *Cephenus* as also attributable to Karsch (1880:656). A check of Karsch's (1880) paper shows that his use of *Cephenus* is a misspelling of Céphène Latreille and not an emendation according to the criteria set forth in the I.C.Z.N. Code.

Ceratolaemus Hesse, 1938:969 (as subgenus of *Platypygus* Loew). Type species: *Platypygus xanthogrammus* Hesse, 1938, by original designation. Synonymized with *Cephalodromia* Becker, 1914, by Bowden (1975b:306).

Chalcamoeba Sack, 1909:522. Type species: *Anthrax virgo* Egger, 1859, by original designation. Synonymized with *Anthrax* Scopoli, 1763, by Becker (1913a:448).

Chalcochiton Loew, 1844:157. Type species: *Chalcochiton speciosus* Loew, 1844, by monotypy. Treated as a subgenus of *Cytherea* Fabricius, 1794, by Engel (1935:321) and Hull (1973:147). **Cythereinae: Cythereini.**

Chasmoneura Hesse, 1938:586. Type species: *Bombylius argyropygus* Wiedemann, 1821, by original designation. Synonymized with *Bombylisoma* Rondani, 1856, by Bowden (1973:150).

Cheilohadrus Hesse, 1938:674. Type species: *Cheilohadrus conspersipennis* Hesse, 1938, by original designation. Synonymized with *Prorachthes* Loew, 1869, by Hesse (1956a:7; 1956b:934).

Chelina Hull, 1973:306 (as subgenus of *Comptosia* Macquart). Type species: *Chelina norrisi* Hull, 1973, by original designation. Synonymized with *Aleucosia* Edwards, 1934, by Yeates (in litt.).

Remarks. Evenhuis (1980:328; 1983b:397; 1989b:366) regarded *Chelina* Hull as an unavailable name because its type species *(norrisi)* was thought to be a *nomen nudum*. Upon subsequent consultation with members of the I.C.Z.N., I have been informed that there is enough information on characters to differentiate the species *norrisi* in the key to genera in Hull (1973:306), thus validating *Chelina* and *norrisi* with Hull as the author of both the genus- and species-group names.

Chiasmella Bezzi, 1924:156. Type species: *Chiasmella brevipennis* Bezzi, 1924, by original designation. Synonymized with *Desmatoneura* Williston, 1895, by Greathead (1988:101).

Remarks. Greathead (1988:101) refers to the mode of type-species fixation as by monotypy, but a check of the original publication shows that it is by original designation (Bezzi 1924:157).

Chionamoeba Sack, 1909:510, 543. Type species: *Bibio nivea* Rossi, 1790, by original designation. Synonymized with *Desmatoneura* Williston, 1895, by François (1967:2).

Choristus Walker, 1852:197. Type species: *Choristus bifrons* Walker, 1852, by monotypy. **Bombyliinae: Dischistini.**

Chrysamoeba Sack, 1909:510, 516. Type species: *Chrysamoeba vulpina* Sack, 1909, by original designation. Junior homonym of *Chrysamoeba* Klebs, 1892. Synonymized with *Argyromoeba* Schiner, 1860, by Becker (1913a:448). Synonymized with *Spogostylum* Macquart, 1840, by Engel (1936a:438).

Chrysanthrax Osten Sacken, 1886b:113 (as subgenus of *Anthrax* Scopoli), 121 (as "group" of *Anthrax*). Type species: *Anthrax fulvohirta* Wiedemann, 1821 [= *Anthrax cypris* Meigen, 1820], by monotypy. Raised to generic status by Hull (1973:365) and again by Hall (1981:2). **Anthracinae: Villini.**

Remarks. Coquillett (1910:523) subsequently designated *Anthrax fulvohirta* Wiedemann, 1821, as the type species of *Chrysanthrax* because he was under the impression that the original description of the genus included more than 1 named species without an original type species designation. Subsequent listings of *Chrysanthrax* have followed this designation (e.g., Painter & Painter 1965; Painter, Painter & Hall 1978). A check of the original description shows that only 1 of the 3 species included by Osten Sacken is named (*fulvohirta* Wiedemann), the other 2 are identified as "may be *Anthrax astarte*" and "perhaps allied to *A. diagonalis.*"

Cillenia: Verrall *in* Scudder, 1882a:74. See *Cyllenia* Latreille.

Cladella Hall and Evenhuis, 1987:619. Type species: *Empidideicus propleuralis*

Melander, 1946, by original designation. Synonym of *Mythenteles* Hall and Evenhuis, 1986, **new synonymy.**

Remarks. Examination of the type species of *Cladella* shows it to fall within the range of variation found in *Mythicomyia mutabilis* Melander, the type species of *Mythenteles* Hall and Evenhuis. Thus, the 2 genera are synonymous, which results in the combination *Mythenteles propleuralis* (Melander), **new combination** and the synonymy *Mythenteles mutabilis* (Melander) = *Mythenteles propleuralis* (Melander), **new synonymy.**

Cladella: Hull, 1973:273. Type species: *Empidideicus propleuralis* Melander, 1946, by original designation. *Nomen nudum.* Objective synonym of *Mythenteles* Hall and Evenhuis, 1986.

Cladodisca Bezzi, 1922a:105 [1922b:143–44; 1923:332; 1924:243] (as subgenus of *Exoprosopa*). Type species: *Exoprosopa munda* Loew, 1869, by monotypy. Synonymized with the subgenus *Heteralonia (Acrodisca)* Bezzi, 1921, by Bowden (1975b:319).

Remarks. Though fully described in Bezzi (1924:243) and dated as such in recent regional catalogs (Bowden 1980:420; Zaitzev 1989:150), the first time *Cladodisca* is made available is in Bezzi (1922a:105) in which the genus-group name is mentioned in association with an available species-group name (*munda* Loew, 1869). Bezzi (1922b:143–44) mentions *Cladodisca* in combination with *Exoprosopa serpentata* Loew, 1854, and later (1923:332; 1924:243) gives *Anthrax suffusa* as the type species. All 3 species (*serpentata, munda,* and *suffusa*) are members of the subgenus *Heteralonia (Acrodisca)* as currently recognized.

Clesthentia White, 1914:45. Type species: *Clesthentia aberrans* White, 1914, by monotypy. ***Incertae sedis* (within non-cyclorrhaphous Diptera).**

Remarks. Michael E. Irwin (pers. comm.) has stated that this genus might be a bombyliid. It was listed provisionally with the Therevidae in Irwin and Lyneborg (1989:358). I have seen the type specimen but am not prepared to make judgment on family placement until more study has been completed.

Codinus: Bigot *in* Scudder, 1882a:79. See *Codionus* Rondani.

Codionus Rondani, 1873:299. Type species: *Codionus chlorizans* Rondani, 1873 [= *Bombylius nobilis* Loew, 1873], by monotypy. Synonymized with *Conophorus* Meigen, 1803, by Paramonov (1929:157).

Remarks. Neave (1939a:784) and Evenhuis (1983b:398) misspelled the name as *Codonius.* The name *Codinus,* attributed to Scudder (1882a) by Schulze et al. (1927:743), is actually an incorrect subsequent spelling of *Codionus* supplied to Scudder by Bigot and not an emendation.

Codonius: Neave, 1939a:784. See *Codionus* Rondani.

Collosoptera Hull, 1973:430. Type species: *Exoprosopa latipennis* Brunetti, 1909, by original designation. **Anthracinae: Exoprosopini.**

Comastes Osten Sacken, 1877:256. Type species: *Comastes robustus* Osten Sacken, 1877, by monotypy. Junior homonym of *Heterostylum* Jan, 1863. Synonymized with *Heterostylum* Macquart, 1848, by Bigot (1892:323).

Remarks. The name *Comastus,* attributed to Wulp (1899:53) by Schulze et al. (1927:767), is merely a misspelling of *Comastes* and not an emendation.

Comastus: Wulp, 1899:53. See *Comastes* Osten Sacken.

Compsella Hull, 1973:354 (as subgenus of *Lyophlaeba* Rondani). Type species: *Macrocondyla transandina* Edwards, 1934, by original designation. Synonymized with the subgenus *Macrocondyla (Lyophlaeba)* Rondani, 1863, by Painter, Painter and Hall (1978:27) [as *Lyophlaeba (Lyophlaeba)*].

Remarks. See also remarks under *Lyophlaeba* Rondani.

Comptosia Macquart, 1840:80 [1841:358]. Type species: *Comptosia fascipennis* Macquart, 1840, by monotypy. **Lomatiinae: Comptosiini.**

Remarks. There are 2 different original spellings of this genus in Macquart (1840) (*Comptosia* on p. 80–81; *Camptosia* on plate 14, fig. 1). I consider *Camptosia* to be an incorrect original spelling of *Comptosia.*

Coniomastix Enderlein, 1934:140. Type species: *Coniomastix montanus* Enderlein, 1934, by original designation. **Nomen dubium** (see Evenhuis 1985a). **Anthracinae: Anthracini.**

Conogaster Hermann, 1907:199. Type species: *Conogaster erythraspis* Hermann, 1907, by original designation. Junior homonym of *Conogaster* Brauer and Bergenstamm, 1891. Synonymized with *Cononedys* Hermann, 1907, by Bezzi (1924:15).

Conomyza Hesse, 1956a:114. Type species: *Conomyza semirufella* Hesse, 1956, by original designation. **Anthracinae: Plesiocerini.**

Cononedys Hermann, 1907:197. Type species: *Anthrax stenura* Loew, 1871, by original designation. **Anthracinae: Aphoebantini.**

Conophorina Becker, 1920:181. Type species: *Conophorina bicellaris* Becker, 1920, by monotypy. **Bombyliinae: Conophorini.**

Remarks. Bowden (1980:398) treats this genus in the Dischistini. I have examined specimens of the type species and find that it is more closely allied to *Conophorus* than *Dischistus,* thus its placement here within the Conophorini.

Conophorus Meigen, 1803:268. Type species: *Bombylius maurus* Mikan, 1796 [= *Bombylius virescens* Fabricius, 1787], by monotypy. **Bombyliinae: Conophorini.**

Coptodicrus Enderlein, 1926:91. Type species: *Coptodicrus vespiformis* Enderlein, 1926, by original designation. Synonymized with the subgenus *Systropus (Diaerops)* Enderlein, 1926, by Bowden (1967:142).

Coptopelma Enderlein, 1926:70. Type species: *Coptopelma schineri* Enderlein, 1926 [misidentification, = *Systropus sanguineus* Bezzi, 1921], by original designation. Synonymized with the subgenus *Systropus (Systropus)* Wiedemann, 1820, by Bowden (1967:144).

Remarks. Enderlein (1926) originally described *Coptopelma* at the generic level for the species *Coptopelma schineri* Enderlein, 1926. *Coptopelma schineri* was proposed (incorrectly) as a new replacement name for *Systropus macilentus* "sensu Schiner (1868) nec Wiedemann" (1820). Schiner never described a species *macilentus* but referred to what he thought was *macilentus* Wiedemann. Bowden (1967) showed that Schiner's (1868) *macilentus* was a misidentification for *sanguineus;* thus, the synonymy of the type species above.

Coquillettia Williston, 1896:65. Type species: *Spogostylum vandykei* Coquillett, 1894, by prior monotypy in Coquillett (1894:433). Junior homonym of *Coquillettia* Uhler, 1890. Synonymized with *Dicranoclista* Bezzi, 1924, implicitly by Bezzi (1924:178) and explicitly by Johnson and Johnson (1959:69).

Remarks. Williston (1896:65) originally proposed *Coquillettia* (incorrectly) as a "nom. nov." for "*Spogostylum* Coquillett *non* Macquart" in his key to North

American genera of Bombyliidae. But Coquillett did not describe *Spogostylum,* he merely used it in a different sense than Macquart. By Williston's (1896:65) back referencing to Coquillett, the type species was thus given by prior monotypy. I have researched Coquillett's papers and the only paper before 1896 in which he used *Spogostylum* is Coquillett (1894), where *vandykei* was the only species listed.

Johnson and Johnson (1959:69) mentioned that *Dicranoclista* and *Coquillettia* were considered congeneric on the advice of R. H. Painter. Though *Coquillettia* is currently considered a synonym of *Dicranoclista* in catalogs (Painter & Painter 1965; Painter, Painter & Hall 1978; Hall & Evenhuis in press), there is some doubt that the 2 are actually congeneric. If *Coquillettia* is ever resurrected from synonymy, a new replacement name will be necessary.

Corsomyza Henning, 1832:329. Type species: *Corsomyza simplex* Henning, 1832 [preoccupied, = *Corsomyza simplex* Wiedemann, 1820], by subsequent designation of Evenhuis (1990a:63). Junior homonym of *Corsomyza* Wiedemann, 1820. Synonymized with *Corsomyza* Wiedemann, 1820, implicitly by Hagen (1862:356) and explicitly by Evenhuis (1990a:63).

Remarks. See remarks under *Apatomyza* Henning.

Corsomyza Wiedemann, 1820:13. Type species: *Corsomyza simplex* Wiedemann, 1820, by original designation (on plate). **Cythereinae: Mariobezziini.**

Corycetta Hull, 1973:413 (as subgenus of *Exoprosopa* Macquart). Type species: *Exoprosopa subfasciata* Engel, 1936, by original designation. Synonymized with *Exoprosopa* Macquart, 1840, by Zaitzev (1989:139).

Coryprosopa Hesse, 1956a:118. Type species: *Coryprosopa lineata* Hesse, 1956, by original designation. **Anthracinae: Plesiocerini.**

Crocidium Loew, 1860a:85 [1860b:267]. Type species: *Crocidium poecilopterum* Loew, 1860, by monotypy. **Bombyliinae: Crocidiini.**

Cryomyia Hull, 1973:131. Type species: *Cryomyia argyrospila* Hull, 1973, by original designation. **Bombyliinae: Bombyliini.**

Cyananthrax Painter, 1959:73. Type species: *Anthrax cyanoptera* Wiedemann, 1830, by original designation. **Anthracinae: Villini.**

Cyclorhynchus Macquart, 1840:114 [1841:392]. Type species: *Cyclorhynchus tes-*

taceus Macquart, 1840, by monotypy. Junior homonym of *Cyclorhynchus* Sundeval, 1835. Objective synonym of *Tmemophlebia* Evenhuis, 1986.

Remarks. Macquart's (1840) original description included 2 different spellings (*Cyclorhynchus* on p. 114; *Cyclorynchus* on p. 123). Evenhuis (1986:19) determined that *Cyclorynchus* was an incorrect original spelling and proposed the new replacement name *Tmemophlebia* for *Cyclorhynchus.*

Cyclorrhynchus Bezzi, 1924:3 (unjustified emendation of *Cyclorhynchus* Macquart). Type species: *Cyclorhynchus testaceus* Macquart, 1840, automatic. Junior homonym of *Cyclorrhynchus* Kaup, 1829. Objective synonym of *Tmemophlebia* Evenhuis, 1986.

Remarks. Scudder (1882a:95) listed *Cyclorrhynchus* (based on a list supplied to him by Verrall), but this is merely a mention of a subsequent misspelling of *Cyclorhynchus* and not an emendation.

Cyclorrhynchus: Verrall in Scudder, 1882a:95. See *Cyclorrhynchus* Bezzi.

Cyclorynchus: Macquart, 1840:123. Incorrect original spelling. See *Cyclorhynchus* Macquart.

Cyllenia Latreille, 1802:429. Type species: *Cyllenia maculata* Latreille, 1802, by monotypy. **Cylleniinae: Cylleniini.**

Remarks. Scudder (1882a:74) lists the name *Cillenia,* but this is merely a record of a subsequent misspelling of *Cyllenia* and not an emendation. Zaitzev (1989:110) erroneously dates *maculata* from Latreille (1805), but the name is validated in the 1802 work.

Cyrthosia: Verrall *in* Scudder, 1882a:98. See *Cyrtosia* Perris.

Cyrtisiopsis Séguy, 1930:80. Type species: *Cyrtisiopsis singularis* Séguy, 1930, by original designation. **Mythicomyiinae.**

Cyrtoides Engel, 1933a:102 (as subgenus of *Empidideicus* Becker). Type species: *Empidideicus efflatouni* Engel, 1933, by monotypy. Raised to generic status by Brues, Melander & Carpenter (1954:337). **Mythicomyiinae.**

Cyrtomorpha White, 1916:185. Type species: *Cyrtomorpha paganica* White, 1916, by monotypy. Junior homonym of *Cyrtomorpha* Graff, 1882. Objective synonym of *Acridophagus* Evenhuis, 1983.

Cyrtomyia Bigot, 1892:340 (new replacement name for *Cyrtophorus* Bigot). Type species: *Cyrtophorus pictipennis* Bigot, 1857, automatic. **Ecliminae.**

Cyrtophorus Bigot, 1857:292. Type species: *Cyrtophorus pictipennis* Bigot, 1857, by monotypy. Junior homonym of *Cyrtophorus* Leconte, 1850. Objective synonym of *Cyrtomyia* Bigot, 1892.

Cyrtosia Perris, 1839:54. Type species: *Cyrtosia marginata* Perris, 1839, by monotypy. **Mythicomyiinae.**

Remarks. Scudder (1882a:98) lists *Cyrthosia,* but this is merely a mention of a subsequent misspelling of *Cyrtosia* and not an emendation.

Cyterea: Verrall *in* Scudder, 1882a:99. See *Cytherea* Fabricius.

Cytherea Fabricius, 1794:413. Type species: *Cytherea obscura* Fabricius, 1794, automatic. **Cythereinae: Cythereini.**

Remarks. The subsequent designation of *Cytherea obscura* Fabricius by Becker (1913a:455) is most often cited as the earliest type designation for *Cytherea;* however, a subsequent type designation for *Mulio* Latreille, 1796, by Latreille (1810) is earlier, and because *Mulio* was proposed as a new replacement name for *Cytherea* (even if unnecessarily), the type species is the same for *Cytherea.* Scudder (1882a:99) lists the name *Cyterea,* but this is merely a mention of a subsequent misspelling of *Cytherea* and not an emendation. There are currently 2 subgenera recognized in this genus based on Hull (1973:147); subg. **Cytherea** s. str. and subg. **Chalcochiton** Loew, 1844.

Dagestania Paramonov, 1929:195. Type species: *Dagestania pusilla* Paramonov, 1929, by monotypy. Synonymized with *Apolysis* Loew, 1844, by Evenhuis (1990a:27).

Dasypalpoides Evenhuis, 1978b:247 (unnecessary new replacement name for *Dasypalpus* Macquart). Type species: *Dasypalpus capensis* Macquart, 1840, automatic. Synonym of *Dasypalpus* Macquart, 1840, **new synonymy.**

Remarks. See remarks under *Dasypalpus* Macquart.

Dasypalpus Macquart, 1840:112 [1841:390]. Type species: *Dasypalpus capensis* Macquart, 1840, by monotypy. **Nomen dubium. Cythereinae: Mariobezziini.**

Remarks. Evenhuis (1978b:247) proposed *Dasypalpoides* as a new replacement name for *Dasypalpus* thinking it was preoccupied by *Dasypalpus* Rafinesque, 1815. Rafinesque's name is a *nomen nudum,* thus the new replacement name was unnecessary. See Evenhuis (1990a:64) for discussion on previous misinterpretation of the type species of this genus and its resultant generic status. Until the type or conspecific specimens of *D. capensis* can be found, it must remain a *nomen dubium.*

Defillipia: Verrall *in* Scudder, 1882a:102. See *Defilippia* Lioy.

Defilippia Lioy, 1864:733. Type species: *Anthrax minos* Meigen, 1804, by subsequent designation of Griffini (1896:42). Synonymized with *Exoprosopa* Macquart, 1840, by Griffini (1896:40).

Remarks. Bezzi (1924:251) gave a later designation of *Anthrax megerlei* Meigen, 1820, as the type of *Defilippia,* stating that it fit better with Lioy's original description of the genus-group name. The reference to Dallas (1865:541) in Scudder (1882a:102) of *Defillipia* merely refers to a misspelling of *Defilippia* and is not an emendation.

Denamyza Hull, 1973:164, 165 (as subgenus of *Corsomyza* Wiedemann). Type species: *Corsomyza ochrostoma* Hesse, 1938, by original designation. Synonymized with *Corsomyza* Wiedemann, 1820, by Bowden (1980:405).

Desmatomyia Williston, 1895:268. Type species: *Desmatomyia anomala* Williston, 1895, by monotypy. **Desmatomyiinae.**

Desmatoneura Williston, 1895:267. Type species: *Desmatoneura argentifrons* Williston, 1895, by monotypy. **Anthracinae: Petrorossiini.**

Deusopora Hull, 1971a:1. Type species: *Deusopora sapphirina* Hull, 1971, by original designation. **Anthracinae: Villini.**

Diaerops Enderlein, 1926:70. Type species: *Systropus marshalli* Bezzi, 1924, by original designation. Reduced to subgeneric status under *Systropus* Wiedemann, 1820, by Bowden (1967:142). **Systropodinae.**

Diatropomma Bowden, 1962:116. Type species: *Diatropomma carcasoni* Bowden, 1962, by original designation. **Anthracinae: Exoprosopini.**

Dicranoclista Bezzi, 1924:178. Type species: *Dicranoclista simpsoni* Bezzi, 1924, by original designation. **Anthracinae: Anthracini.**

Remarks. See remarks under *Coquillettia* Williston.

Dimelopelma Enderlein, 1926:90. Type species: *Dimelopelma tessmanni* Enderlein, 1926, by original designation. Reduced to subgeneric status under *Systropus* Wiedemann, 1820, by Bowden (1967:130). **Systropodinae.**

Dimorpaphus: Bigot *in* Scudder, 1882a:108. See *Dimorphaphus* Walker.

Dimorphaphus: Walker, 1871a:255. Type species: *Dimorphophora syrphoides* Walker, 1871, by present designation of 1st species included herein. *Nomen nudum.* Lapsus for *Dimorphophora* Walker, 1871.

Remarks. Dimorpaphus, listed in Schulze et al. (1928:1002) and by Scudder (1882a:108), is merely a misspelling of *Dimorphaphus* in a list of Diptera supplied to Scudder by Bigot and is not an emendation.

Dimorphophora Walker, 1871b:272. Type species: *Dimorphophora syrphoides* Walker, 1871 [= *Antonia suavissima* Loew, 1856], by monotypy. Synonymized with *Antonia* Loew, 1844, by Loew (1873:180).

Diochanthrax Hall, 1975a:109 (as subgenus of *Villa* Lioy). Type species: *Villa morulus* Hall, 1975, by monotypy. Raised to generic status by Hall (1981:2). **Anthracinae: Villini.**

Dipalta Osten Sacken, 1877:236. Type species: *Dipalta serpentina* Osten Sacken, 1877, by monotypy. **Anthracinae: Villini.**

Diplocampta Schiner, 1867:312 [1868:119]. Type species: *Diplocampta singularis* Schiner, 1867, by original designation. **Anthracinae: Villini.**

Dischistus Loew, 1855:45 (as subgenus of *Bombylius*). Type species: *Bombylius mystax* Wiedemann, 1818, by subsequent designation of Becker (1913a:494). Raised to generic status by Bigot (1858:584). **Bombyliinae: Dischistini.**

Remarks. Bowden (1973) discusses the confusion over the generic concept of this name based on 2 different type species being designated: *Bombylius mystax* Wiedemann, 1818 by Becker (1913a:494) and *Bombylius minimus* Schrank, 1781, by Brunetti (1920:273). Many authors have followed Brunetti's (1920) designation, but Becker's (1913a) is the earlier of the two.

In his "Notanda corrigenda et mutunda," Rondani (1861:8) makes back-reference to his 1856 work on p. 164 and states "pro *Bombylisoma* R.—pone *Dischistus* Loëw 1855." Since Rondani (1856:164) gives *Bombylius sulphureus* as the type species for *Bombylisoma,* the substitution of *Dischistus* for *Bombylisoma* makes sulphureus the type-species for *Dischistus* based on the earliest subsequent designation in Rondani (1861:8) (see Evenhuis 1991b for details). Because acceptance of the subsequent type designation of *sulphureus* for *Dischistus* would cause instability of taxonomy and common usage, an application to the I.C.Z.N. has been submitted to suppress all type designations for *Dischistus* previous to Becker (1913a).

Docidomyia White, 1916:200, 203. Type species: *Docidomyia puellaris* White, 1916, by monotypy. **Tomomyzinae.**

Remarks. Hull (1973) placed *Docidomyia* in his tribe Myonematini within the Lomatiinae, which was followed by Evenhuis (1989b:368). This placement is incorrect because *Docidomyia* possesses characters that place it with genera allied to *Paracosmus* in the subfamily Tomomyzinae.

Doddosia Edwards, 1934:101. Type species: *Doddosia picta* Edwards, 1934, by original designation. **Lomatiinae: Comptosiini.**

Dolichomyia Wiedemann, 1830:642. Type species: *Dolichomyia nigra* Wiedemann, 1830, by monotypy. **Systropodinae.**

Doliogethes Hesse, 1938:545. Type species: *Bombylius seriatus* Wiedemann, 1821, by original designation. **Bombyliinae: Dischistini.**

Doliopteryx Hesse, 1938:39, 968 [1956a:15 (key)]. Type species: *Doliopteryx crocea* Hesse, 1956, by present designation of 1st species included herein. Unavailable name. Proposed after 1930 without a type species. Objective synonym of *Doliopteryx* Hesse, 1956.

Remarks. Hesse (1938:39, 968) mentioned that this genus would be published in the 2nd part (Hesse 1956a,b) of his revision of southern African Bombyliidae, but it is listed without a designated type species. It is diagnosed in a key to genera in Hesse (1956a:15), but again without included species or a type species designation.

Doliopteryx Hesse, 1956b:936. Type species: *Doliopteryx crocea* Hesse, 1956, by original designation. **Mythicomyiinae.**

Dumontiella Séguy, 1932:126. Type species: *Dumontiella vespertilio* Séguy, 1926, by original designation. Synonymized with *Prorachthes* Loew, 1869, by Paramonov (1934b:109).

Eccremus: Wiedemann, 1821:243. Type species: *Phthiria albida* Wiedemann, 1821, by monotypy. Unavailable name. Originally proposed in synonymy with *Phthiria* Meigen; not made available before 1961. Synonym of *Tmemophlebia* Evenhuis, 1986, **new synonymy.**

Remarks. Wiedemann (1821:242) described *Phthiria albida* from Brazil as a new species. In his remarks under the species description he stated that the specimen on which he based the description was under the name *Eccremus albidus* in Hoffmansegg's collection. Scudder (1882a:117) and Neave (1939b:180) listed *Eccremus* as attributed to Wiedemann, 1821, but without pagination or mention of its status in nomenclature. *Phthiria albida* is currently a member of the genus *Tmemophlebia,* hence *Eccremus* must be placed in synonymy with this genus.

Eclimmus Agassiz, 1847:134(unjustified emendation of *Eclimus* Loew). Type species: *Eclimus perspicillaris* Loew, 1844, automatic. Synonymized with *Eclimus* Loew, 1844, by Bezzi (1903:188).

Eclimus Loew, 1844:154. Type species: *Eclimus perspicillaris* Loew, 1844, by subsequent designation of Coquillett (1910:536). **Ecliminae.**

Ecliptica: Engel, 1933a:103. Type species: *Empidideicus efflatouni* Engel, 1933, by monotypy. Unavailable name. Originally proposed in synonymy with the subgenus *Empidideicus (Cyrtoides)* Engel, 1933; not made available before 1961.

Edmundiella Becker, 1915:347. Type species: *Edmundiella niveifrons* Becker, 1915, by monotypy. **Lomatiinae: Lomatiini.**

Efflatounia Bezzi, 1925:168, 169, 170. Type species: *Efflatounia aegyptiaca* Bezzi, 1925, by original designation. **Bombyliinae: Dischistini.**

Remarks. Bowden (1973:144) synonymized *Karakumia* under *Efflatounia* but later (1985:99) resurrected it from synonymy and placed *Neobombylodes* Evenhuis, 1978 (along with *Bombylodes* Paramonov, 1940) in synonymy with *Efflatounia* based on some characters that have, upon further study, proven symplesiomorphic. The genera *Efflatounia, Karakumia,* and *Neobombylodes* indeed form a close group, but I have found autapomorphies that distinguish each genus. I prefer to follow Zaitzev (1989) in treating all 3 as distinct genera.

Elachymyia Hall and Evenhuis, 1987:616 (as subgenus of *Paraconsors* Hall and Evenhuis). Type species: *Empidideicus humeralis* Melander, 1946, by original designation. **Mythicomyiinae.**

Empidideicus Becker, 1907a:97. Type species: *Empidideicus carthaginiensis* Becker, 1907, by monotypy. **Mythicomyiinae.**

Empidigeron Painter, 1932:143 (as subgenus of *Geron* Meigen). Type species: *Geron snowi* Painter, 1932, by original designation. **Gerontinae.**

Enica Macquart, 1834:399. Type species: *Anthrax longirostris* Wiedemann, 1819, by monotypy. **Cylleniinae: Enicini.**

Eniconeura: Agassiz, 1846:13. See *Eniconevra* Macquart.

Eniconevra Macquart, 1840:110 [1841:388]. Type species: *Eniconevra fuscipennis* Macquart, 1840, by monotypy. Synonymized with *Toxophora* Meigen, 1803, by Kertész (1909:91) and Verrall (1909a:510; 1909b:17).

Remarks. As listed in nomenclators (e.g., Scudder 1882a:123; Neave 1939b:235), Agassiz (1846:13) used the spelling *Eniconeura.* This is merely an incorrect subsequent spelling (also carried over to Agassiz's 1847 work) and does not satisfy the criteria to be an emendation.

Epacmoides Hesse, 1956a:125. Type species: *Plesiocera biumbonata* Bezzi, 1922, by original designation. **Anthracinae: Plesiocerini.**

Epacmus Osten Sacken, 1886a:76 (November) [1887:142 (January) (new replacement name for *Leptochilus* Loew)]. Type species: *Leptochilus modestus* Loew, 1872, by subsequent monotypy in Osten Sacken (1887:142). **Anthracinae: Aphoebantini.**

Remarks. The fact that the *Biologia Centrali-Americana* appeared in fascicles apparently eluded previous catalogers (see also under *Astrophanes* Osten Sacken) and results in an earlier date of validation for *Epacmus* than previously recorded and an unusual circumstance regarding the type species fixation for this genus. By mentioning *Epacmus* in the key to genera (Osten Sacken 1886a:76 [issued in November 1886]), the name is valid from that date but has no species included. Because it was not specifically proposed as a new replacement name in the key, it requires a subsequent fixation of its type species either by subsequent monotypy or by subsequent designation of the 1st species included in the genus. Osten Sacken (1887:142 [issued in January 1887]) again proposes the name *Epacmus,* but this time as a new replacement name for the preoccupied *Leptochilus* Loew, 1872. On the same page, Osten Sacken (1887:142) includes the 1st and only species for *Epacmus* (*Leptochilus modestus* Loew, 1872). Thus, the type species for *Epacmus* Osten Sacken, 1886, is by subsequent monotypy.

Epibates Osten Sacken, 1877:268. Type species: *Epibates funestus* Osten Sacken, 1877, by subsequent designation of Coquillett (1910:538). Synonymized with *Thevenemyia* Bigot, 1875, by Bigot (1887b:lx; 1887d:lx). Synonymized with *Thevenetimyia* Bigot, 1875, by Bigot (1892:325).

Epidosia Hull, 1973:306 (as subgenus of *Comptosia* Macquart). Type species: *Comptosia vittata* Edwards, 1934, by original designation. Synonymized with *Comptosia* Macquart, 1840, by Evenhuis (1989b:367).

Remarks. Hull (1973) used 2 different original spellings for this genus (*Epidosia* on p. 306; *Epidosina* on p. 349, 658). I consider *Epidosina* an incorrect original spelling of *Episodia* Hull, 1973.

Epidosina: Hull, 1973:349, 658. Incorrect original spelling. See *Epidosia* Hull.

Eristalopsis Evenhuis, 1985b:289 (new replacement name for *Syrphoides* Evenhuis). Type species: *Bombylius byrrhus* Bowden, 1971, automatic. **Bombyliinae: Dischistini.**

Euanthobates Hesse, 1965:482. Type species: *Euanthobates pectinigulus* Hesse, 1965, by original designation. **Mythicomyiinae.**

Eucessia Coquillett, 1886a:82. Type species: *Eucessia rubens* Coquillett, 1886, by monotypy. **Anthracinae: Aphoebantini.**

Eucharimyia Bigot, 1889 (8 January):cxl. Type species: *Eucharimyia dives* Bigot, 1889 [preoccupied, = *Euchariomyia dives* Bigot, 1888], by monotypy. Synonym of *Euchariomyia* Bigot, 1888, **new synonymy.**

Remarks. See remarks under *Euchariomyia* Bigot.

Euchariomyia Bigot, 1888:cxl (6 October). Type species: *Euchariomyia dives* Bigot, 1888, by monotypy. **Bombyliinae: Bombyliini.**

Remarks. Rediscovery of the *Bull. Bimens. Soc. Entomol. Fr.* (see Evenhuis et al. 1989:810), an earlier, and in some cases different, version of the *Bulletin* bound with the *Annales,* has led to earlier dates of publication and, in some cases, different spellings for genera and species described by Bigot and other authors. In this particular case, *Euchariomyia* is the 1st valid spelling for this genus and dates from 1888. Neave's (1939b:309) listing of *Euchariomyia* as a misspelling of *Eucharimyia* and attributable to Wulp (1896:73) is not completely correct. Wulp (1896) was, in fact, citing the 1st valid spelling of the genus *(Euchariomyia)* giving the correct dating (as 1888) and mentions in a footnote that Bigot changed the name to *Eucharimyia* "in litt." The accepted spelling here of *Euchariomyia,* though different from traditional usage, is so slight as to hardly cause confusion among bombyliid workers.

Euprepina Hull, 1971b:181. Type species: *Euprepina nuda* Hull, 1971, by original designation. **Bombyliinae: Dischistini.**

Euricarenus: Verrall *in* Scudder, 1882a:133. See *Eurycarenus* Loew.

Eurycarena: Marschall, 1873:331. See *Eurycarenus* Loew.

Eurycarenus Loew, 1860a:83 [1860b:258]. Type species: *Bombylius laticeps* Loew, 1852, by monotypy. **Bombyliinae: Bombyliini.**

Remarks. Scudder (1882b:121) and Schulze et al. (1929:1262) listed *Eurycarena* as attributable to Marschall (1873:331). I have not been able to verify the entry in Marschall (1873) but assume that this is merely a misspelling for *Eurycarenus* and not an emendation. Based on a list supplied to him by Verrall, Scudder (1882a:133) listed *Euricarenus,* but this is merely a mention of a subsequent spelling of *Eurycarenus* and not an emendation.

Euryphthiria Evenhuis, 1986:9. Type species: *Euryphthiria grandis* Evenhuis, 1986, by original designation. **Phthiriinae: Poecilognathini.**

Eusurbus Roberts, 1929:580. Type species: *Bombylius crassilabris* Macquart, 1855, by original designation. **Bombyliinae: Dischistini.**

Exechohypopion Evenhuis, 1991a:283. Type species: *Exechohypopion greatheadi* Evenhuis, 1991, by original designation. **Anthracinae: Villini.**

Exepacmus Coquillett, 1894:101. Type species: *Exepacmus johnsoni* Coquillett, 1894, by monotypy. **Anthracinae: Aphoebantini.**

Exhyalanthrax Becker, 1916:44 (as subgenus of *Villa* Lioy). Type species: *Anthrax* *vagans* Loew, 1862, by subsequent designation of Bezzi (1924:7, 192). Raised to generic status by Bowden (1980:414). **Anthracinae: Villini.**

Remarks. Theodor (1983:241) returned *Exhyalanthrax* to subgeneric status under *Thyridanthrax,* but I prefer to keep it as a separate genus, the treatment of which has been followed by more recent works (e.g., Greathead 1988a:106; Evenhuis 1989b:371; Zaitzev 1989:136).

Exoprosopa Macquart, 1840:35 [1841:313]. Type species: *Anthrax pandora* Fabricius, 1805, by subsequent designation of Coquillett (1910:544). **Anthracinae: Exoprosopini.**

Remarks. Rondani (1856:162) designated *Bibio capucina* Fabricius, 1782, as the type species, which was followed by some subsequent workers and regional catalogs (e.g., Lioy 1864; Evenhuis 1989b). However, *Bibio capucina* was not an originally included species, thus Rondani's (1856) designation is not valid. Coquillett's (1910) subsequent designation of *Anthrax pandora* as the type species of *Exoprosopa* has been followed by most recent workers (e.g., Painter & Painter 1965; Bowden 1980; Zaitzev 1989). The earliest valid designation, though, is by Duponchel in d'Orbigny (1845:545) of *Exoprosopa audouinii* Macquart, 1840 (see Evenhuis & Thompson 1990:235 for more details). Acceptance of this type species would make *Exoprosopa* a senior synonym of *Ligyra* Newman, 1841, and would upset taxonomic and nomenclatural stability in the subfamily. Evenhuis and Greathead (1989:250) have applied to the I.C.Z.N. to conserve *Anthrax pandora* Fabricius, 1805, as the type species of *Exoprosopa* and to ignore all type designations of *Exoprosopa* previous to Coquillett (1910).

Sherborn (1922–1932, 1932–1933) listed some of Macquart's (1840) species as being proposed under the genus-group name *Exoprosopis.* A check of the text, plates, and plate legends of the journal and reprint versions of Macquart's work shows that only the spelling *Exoprosopa* was used. Sherborn's *Exoprosopis* is thus a lapsus.

Exoptata Coquillett, 1887:13 (as subgenus of *Exoprosopa* Macquart). Type species: *Exoprosopa divisa* Coquillett, 1887, by monotypy. Synonymized with *Exoprosopa* Macquart, 1840, by Kertész (1909:5) and Verrall (1909b:18).

Exosoma: Séguy, 1938:333. Type species: *Exosoma erythraea* Séguy, 1938 [= *Tanyglossa maurus* Thunberg, 1827], by monotypy. Unavailable name. Proposed in synonymy with *Litorhynchus maurus* Thunberg; not made available before 1961. Junior homonym of *Exosoma* Jacoby, 1903. Synonymized with *Litorhina* Bowden, 1975, by Bowden (1975b:317).

Francoisia Hesse, 1952:1 (as genus). Type species *Francoisia sulcifacies* Hesse, 1952 [= *Isotamia daveyi* Bezzi, 1912], by original designation. Synonymized with *Isotamia* Bezzi, 1912, by François (1954:180). *Isotamia* treated as a subgenus of *Heteralonia* Rondani, 1863, by Bowden (1980:422).

Geminaria Coquillett, 1894:109. Type species: *Lordotus canalis* Coquillett, 1887, by original designation. **Bombyliinae: Dischistini.**

Geron Meigen, 1820:223. Type species: *Geron gibbosus* Meigen, 1820 [preoccupied, = *Bombylius gibbosus* Olivier, 1789], by subsequent designation of Duponchel *in* d'Orbigny (1845:206). **Gerontinae.**

Remarks. Current catalogs (Painter & Painter 1965:423; Bowden 1980:385; Evenhuis 1989b:361) recognize 4 subgenera in this genus: subg. **Geron** s. str., subg. **Empidigeron** Painter, 1932, subg. **Plichtamyia** Evenhuis, 1979, and subg. **Pseudoammictus** Bigot, 1892.

Geronites† Cockerell, 1915:643. Type species: *Geronites stigmalis* Cockerell, 1915, by monotypy. Synonymized with *Poecilognathus* Jaennicke, 1867, by Evenhuis (1990a:53). Fossil shale genus.

Glabella Loew, 1873:210. Type species: *Glabella femorata* Loew, 1872, by monotypy. Junior homonym of *Glabella* Swainson, 1840. Synonymized with *Glabellula* Bezzi, 1902, by Bezzi (1902:191).

Remarks. Glabella Loew, 1873, placed on the Official Index of Rejected and Invalid Generic Names in Zoology by the I.C.Z.N. (1989:148).

Glabellula Bezzi, 1902:191 (new replacement name for "*Platygaster* . . . und *Sphaerogaster* . . . und *Glabella*"). Type species: *Platygaster arcticus* Zetterstedt, 1838, by subsequent designation of I.C.Z.N. (1989:148). **Mythicomyiinae.**

Remarks. Glabellula Bezzi, 1902, placed on the Official List of Generic Names in Zoology by I.C.Z.N. (1989:148).

Glaesamictus† Hennig, 1966:12. Type species: *Glaesamictus hafniensis* Hennig, 1966, by monotypy. Baltic amber genus. **Cylleniinae: Cylleniini.**

Glossista Rondani, 1856:163. Type species: *Mulio infuscatus* Meigen, 1820, by original designation. Synonymized with *Mulio* Latreille, 1796, by Loew (1860b:248). Synonymized with *Cytherea* Fabricius, 1794, by Kertész (1909:70).

Glycophorba Hull, 1973:295 (as subgenus of *Amphicosmus* Coquillett). Type species: *Amphicosmus cincturus* Williston, 1901, by original designation. **Tomomyzinae.**

Gnumya: Bezzi, 1921a:173, 177. Incorrect original spelling. See *Gnumyia* Bezzi.

Gnumyia Bezzi, 1921a:82. Type species: *Gnumyia brevirostris* Bezzi, 1921, by original designation. **Cythereinae: Mariobezziini.**

Remarks. Two different spellings of this genus appear in the original publication (*Gnumyia* on p. 4, 82 and *Gnumya* on p. 173, 177). I consider *Gnumya* to be an incorrect original spelling of *Gnumyia* Bezzi, 1921.

Gonarthrus Bezzi, 1921a:88 [1923:324; 1924:109]. Type species: *Dischistus cylindricus* Bezzi, 1906, by original designation. **Bombyliinae: Dischistini.**

Gyrocraspedum Becker, 1913a:456. Type species: *Gyrocraspedum pleskei* Becker, 1913, by original designation. **Cythereinae: *Tribus inquirendus.***

Hallidia Hull, 1970:165. Type species: *Hallidia plumipilosa* Hull, 1970, by original designation. **Bombyliinae: Dischistini.**

Hemipenthes Loew, 1869b:28 [1872b:141]. Type species: *Musca morio* Linnaeus, 1758, by subsequent designation of Coquillett (1910:550). **Anthracinae: Villini.**

Remarks. See *Anthrax* Scopoli for information on type species conflict in *Anthrax* and *Hemipenthes.*

Henica Agassiz, 1847:138, 178 (unjustified emendation of *Enica* Macquart). Type species: *Anthrax longirostris* Wiedemann, 1819, automatic. Objective synonym of *Enica* Macquart, 1834.

Heniconeura Bezzi, 1903:189 (unjustified emendation of *Eniconevra* Macquart). Type species: *Eniconevra fuscipennis* Macquart, 1840, automatic. Synonym of *Toxophora* Meigen, 1803.

Heniconeura: Verrall *in* Scudder, 1882:159. See *Heniconevra* Agassiz.

Heniconevra Agassiz, 1847:138, 178 (unjustified emendation of *Eniconevra* Macquart). Type species: *Eniconevra fuscipennis* Macquart, 1840, automatic. Synonym of *Toxophora* Meigen, 1803.

Remarks. Scudder (1882a:159) lists *Heniconeura* and attributes it to Verrall. *Heniconeura* is merely a listing of a subsequent misspelling of *Eniconevra* and not an emendation.

Heteralonia Rondani, 1863:57 [1864:57]. Type species: *Exoprosopa oculata* Macquart, 1840 (as "*occulta*"), by monotypy. **Anthracinae: Exoprosopini.**

Remarks. According to Bowden (1975b:318–19; 1980:420–22), there are 6 subgenera recognized in this genus: subg. **Heteralonia** s. str., subg. **Acrodisca** Bezzi, 1921, subg. **Homolonia** Bowden, 1975, subg. **Isotamia** Bezzi, 1912, subg. **Mesoclis** Bezzi, 1921, and subg. **Metapenta** Bezzi, 1921. See also remarks under *Mima* Meigen.

Heterhybos Brèthes, 1920:40. Type species: *Heterhybos hyalinipennis* Brèthes, 1920, by original designation. Synonymized with *Mythicomyia* Coquillett, 1893, by Melander (1928:375).

Heterostilum: Macquart, 1848:231(71). Incorrect original spelling. See *Heterostylum* Macquart.

Heterostylum Macquart, 1848:195(35). Type species: *Heterostylum flavum* Macquart, 1848, by original designation. **Bombyliinae: Dischistini.**

Remarks. Macquart used 2 different spellings of this name in the original publication [*Heterostylum* on p. 195(35) and *Heterostilum* on p. 231(71)]. I consider *Heterostilum* to be an incorrect original spelling of *Heterostylum.*

Heterotropus Loew, 1873:182. Type species: *Heterotropus albidipennis* Loew, 1873, by monotypy. **Heterotropinae.**

Hilarimorpha Schiner, 1860a:54. Type species: *Hilarimorpha singularis* Schiner, 1860, by original designation. ***Incertae sedis* (within non-cyclorrhaphous Diptera).**

Remarks. This genus has been bounced from family to family within the orthorrhaphous Brachycera and even placed in its own family, Hilarimorphidae. Woodley (1989:1384) returned it to the Bombyliidae as a result of his phylogenetic analysis of Nearctic orthorrhaphous Brachycera families. Other more recent phylogenetic studies on related families (Wiegmann 1989) show different results.

Homolonia Bowden, 1975b:319 (as subgenus of *Heteralonia* Rondani). Type species: *Anthrax megerlei* Meigen, 1820, by original designation. **Anthracinae: Exoprosopini.**

Remarks. See also remarks under *Mima* Meigen.

Hyalanthrax Osten Sacken, 1886b:112 (December) [1887:134 (January)] (as subgenus of *Anthrax* Scopoli). Type species: *Anthrax faustinus* Osten Sacken, 1887, by subsequent designation of Coquillett (1910:553). Synonymized with *Villa* Lioy, 1864, by Brues and Melander (1932:295).

Remarks. Works listing *Hyalanthrax* commonly date it from Osten Sacken (1887:134) (e.g., Painter & Painter 1965:433; Hall 1975b:212; Painter, Painter & Hall 1978:45). However, Osten Sacken (1886b:112) gives characters validating *Hyalanthrax* without including any species. Coquillett's (1910:553) designation of *Anthrax faustinus* is from the 1st inclusion of 8 named species in the subgenus by Osten Sacken (1887:134).

Hyperalonia Rondani, 1863:58 [1864:58]. Type species: *Anthrax erythrocephala* Fabricius, 1805, by subsequent designation of Dallas (1866:649). Treated as a subgenus of *Ligyra* Newman, 1841, by Theodor (1983:233). **Anthracinae: Exoprosopini.**

Remarks. Previous listings of this genus commonly give the subsequent type designation of Coquillett (1910:554) of the same species, but Dallas (1866:649) is earlier. A later type designation by Brunetti (1920:178) is also of the same species.

Hyperusia Bezzi, 1921a:84 [1924:107]. Type species: *Hyperusia minor* Bezzi, 1921, by present designation. **Cythereinae: Mariobezziini.**

Remarks. The original publication includes 2 species (*minor* on p. 84 and *soror* on p. 167) without a type-species designation. Bezzi (1924:107) also described it as a new genus and gave *Hyperusia luteifacies* Bezzi, 1924, as the type species by original designation. This designation is invalid, however, because *lutiefacies* was not one of the original species included. No valid subsequent designations of a type species have been made previously for this genus, thus the present designation.

Inyo Hall and Evenhuis, 1987:638. Type species: *Heterotropus senex* Melander, 1950, by original designation. **Desmatomyiinae.**

Isocnemus Bezzi, 1924:101. Type species: *Isocnemus nemestrinus* Bezzi, 1924, by original designation. **Bombyliinae: Dischistini.**

Isopenthes Osten Sacken, 1886a:96. Type species: *Isopenthes jaennickeana* Osten Sacken, 1886, by subsequent designation of Coquillett (1910:556). Synonymized with *Hemipenthes* Loew, 1869, by Painter (1933a:5).

Isotamia Bezzi, 1912:606, 627. Type species: *Isotamia daveyi* Bezzi, 1912, by original designation. Treated as a subgenus of *Heteralonia* Rondani, 1863, by Bowden (1980:422). **Anthracinae: Exoprosopini.**

Karakumia Paramonov, 1927:77. Type species: *Karakumia nigra* Paramonov, 1927, by monotypy. **Bombyliinae: Dischistini.**

Remarks. The date of the original description of *Karakumia* is usually given as 1926. I follow Paramonov (1939b:35) in dating it from 1927. See also remarks under *Efflatounia* Bezzi.

Lagochilus Loew, 1860a:87 [1860b:273]. Type species: *Cyllenia afra* Wiedemann, 1828 [= *Anthrax longirostris* Wiedemann, 1819], by monotypy. Synonymized with *Enica* Macquart, 1840, by Schiner (1868:115).

Laminanthrax Greathead, 1967:251. Type species: *Laminanthrax albus* Greathead, 1967 [= *Thyridanthrax chionophorus* Bezzi, 1925], by original designation. **Anthracinae: Villini.**

Lasioprosopa Macquart, 1855:102(82). Type species: *Lasioprosopa bigotii* Macquart, 1855 [= *Corsomyza nigripes* Wiedemann, 1820], by monotypy. Synonymized with *Corsomyza* Wiedemann, 1820, by Bezzi (1924:13).

Laurella Hull, 1971a:3. Type species: *Laurella auripila* Hull, 1971 [= *Dischistus formosus* Roberts, 1928], by original designation. **Bombyliinae: Dischistini.**

Legnotomyia Bezzi, 1902:191 (new replacement name for *Legnotus* Loew). Type species: *Bombylius trichorhoeus* Loew, 1855, automatic. **Bombyliinae: Conophorini.**

Legnotus Loew, 1855:6, 41 (as subgenus of *Bombylius*). Type species: *Bombylius trichorhoeus* Loew, 1855, by original designation. Raised to generic status by Bigot (1858:584). Junior homonym of *Legnotus* Schiödte, 1848. Objective synonym of *Legnotomyia* Bezzi, 1902.

Remarks. Schulze et al. (1931:1784) and Neave (1939b:885) listed *Legonotus* as attributable to Bischof (1903:42). A check of Bischof's article shows that *Legonotus* is merely a misspelling of *Legnotus,* as correctly cited by Kertész (1909:109) and is not an emendation or proposal of a new genus-group name.

Legonotus: Bischof, 1903:42. See *Legnotus* Loew.

Lepidanthrax Osten Sacken, 1886a:76 (November) [1886b:107 (December)]. Type species: *Anthrax disjunctus* Wiedemann, 1830, by subsequent designation of Coquillett (1910:559). **Anthracinae: Villini.**

Remarks. The earliest valid publication of *Lepidanthrax* is from the key to genera in Osten Sacken (1886a:76) with no species included. Coquillett's designation of *Anthrax disjunctus* is from the 1st included 3 species included in Osten Sacken (1886b:107).

Lepidochlanus Hesse, 1938:613. Type species: *Lepidochlanus fimbriatus* Hesse, 1938, by original designation. **Bombyliinae: Dischistini.**

Lepidophora Westwood, 1835:447. Type species: *Ploas aegeriiformis* Gray, 1832 [= *Toxophora lepidocera* Wiedemann, 1828], by monotypy. **Ecliminae.**

Leptochilus Loew, 1872a:78 [1872b:254]. Type species: *Leptochilus modestus* Loew, 1872, by monotypy. Junior homonym of *Leptochilus* Saussure, 1852. Objective synonym of *Epacmus* Osten Sacken, 1886.

Leucamoeba Sack, 1909:520. Type species: *Bibio aethiops* Fabricius, 1781, by original designation. Synonymized with *Spogostylum* Macquart, 1840, by Bezzi (1924:15).

Leylaiya Efflatoun, 1945:21. Type species: *Leylaiya mimnermia* Efflatoun, 1945, by original designation. **Mythicomyiinae.**

Ligyra Newman, 1841:220. Type species: *Anthrax bombyliformis* Macleay, 1826, by monotypy. **Anthracinae: Exoprosopini.**

Remarks. The name *Lygira* listed by Scudder (1882a:196) is merely a listing of a subsequent misspelling of *Ligyra* and not an emendation.

According to Theodor (1983:233) there are 2 subgenera currently recognized in this genus: subg. **Ligyra** s. str. and subg. **Hyperalonia** Rondani, 1863.

Liophleba Bezzi, 1924:4 (unjustified emendation of *Lyophlaeba* Rondani). Type species: *Lyophlaeba lugubris* Rondani, 1863, automatic. Objective synonym of *Lyophlaeba* Rondani, 1863.

Lissomerus Austen, 1937:33. Type species: *Lissomerus niveicomatus* Austen, 1937, by original designation. Synonymized with *Dischistus* Loew, 1855, by Bowden (1973:140).

Lithocosmus† Cockerell, 1909c:71. Type species: *Lithocosmus coquilletti* Cockerell, 1909, by monotypy. Fossil shale genus. **Cylleniinae: *Tribus inquirendus.***

Litomyza: Hull, 1973:426. Type species: *Litorhynchus hamatus* Macquart, 1840, by present designation. *Nomen nudum.* Synonym of *Exoprosopa* Macquart, 1840.

Litorhina Bowden, 1975:314. Type species: *Litorhynchus corticeus* Bezzi, 1924, by original designation. **Anthracinae: Exoprosopini.**

Litorhynchus Macquart, 1840:78 [1841:356]. Type species: *Litorhynchus hamatus* Macquart, 1840 [= *Anthrax heros* Wiedemann, 1819], by subsequent designation of Coquillett (1910:562). Synonymized with *Exoprosopa* Macquart, 1840, originally by Loew (1860b:248) and again by Bowden (1975b:316) after clarification of generic concepts.

Remarks. Bezzi (1924:3) also gives *hamatus* as the type species of *Litorhynchus* Macquart by subsequent designation. Even later, Oldroyd (1940:202), thinking the genus had no type species, gave *Bibio lar* Fabricius, 1781, as the type species by subsequent designation. See Bowden (1975b:316) for a detailed discussion of the synonymic history of this genus.

Litorrhynchus Verrall *in* Scudder, 1882a:192 (unjustified emendation of *Litorhynchus* Macquart). Type species: *Litorhynchus hamatus* Macquart, 1840, automatic. Synonymized with *Litorhynchus* Macquart, 1840, by Hull (1973:426). Synonymized with *Exoprosopa* Macquart, 1840, herein, **new synonymy.**

Remarks. Litorrhynchus Bezzi (1924:3) is a later unjustified emendation of *Litorhynchus.*

Logcocerius Rondani, 1863:61 [1864:61]. Type species: *Anthrax holosericeus* Fabricius, 1794 (as "*M. holosericeus* Loew"), by original designation. Synonymized with *Cytherea* Fabricius, 1794, by Kertész (1909:70).

Remarks. Scudder (1882a:193) lists *Logiocerius,* but this is merely a listing of a subsequent misspelling of *Logcocerius* and is not an emendation.

Logiocerius: Verrall *in* Scudder, 1882a:193. See *Logcocerius* Rondani.

Lomatia Meigen, 1822:[v] [1830:324] (new replacement name for *Stygia* Meigen). Type species: *Anthrax belzebul* Fabricius, 1794, by subsequent designation of Westwood (1840:131). **Lomatiinae: Lomatiini.**

Lonchocerius Bezzi, 1924:4 (unjustified emendation of *Logcocerius* Rondani). Type species: *Anthrax holosericeus* Fabricius, 1794, automatic. Synonymized with *Cytherea* Fabricius, 1794, by Hull (1973:145).

Loncocerius Bezzi, 1903:182 (unjustified emendation of *Logcocerius* Rondani). Type species: *Anthrax holosericeus* Fabricius, 1794, automatic. Synonymized with *Cytherea* Fabricius, 1794, by Hull (1973:145).

Lordotus Loew, 1863:303 [1864:187]. Type species: *Lordotus gibbus* Loew, 1863, by monotypy. **Bombyliinae: Dischistini.**

Lygira: Verrall *in* Scudder, 1882a:196. See *Ligyra* Newman.

Lyophlaeba Rondani, 1863:54, 55 [1864:54, 55]. Type species: *Lyophlaeba lugubris* Rondani, 1863, by original designation. Treated as a subgenus of *Macrocondyla* Rondani, 1863, by Edwards (1934:103). **Lomatiinae: Comptosiini.**

Remarks. Hall (1973:149) treated *Lyophlaeba* as the nominate subgenus for it and *Macrocondyla* stating page priority as the reason for doing so. Edwards (1934) was the 1st revisor of the 2 names according to the criteria set forth in Art. 24 of the I.C.Z.N. Code and treated *Macrocondyla* as the nominate subgenus. Edwards' (1934) treatment is the action to follow rather than page priority.

Macrocondyla Rondani, 1863:54, 55 [1864:54, 55]. Type species: *Macrocondyla pictinervis* Rondani, 1863, by original designation. **Lomatiinae: Comptosiini.**

Remarks. According to Edwards (1934), there are 2 subgenera recognized in this genus: subg. **Macrocondyla** s. str. and subg. **Lyophlaeba** Rondani, 1863. See also remarks under *Lyophlaeba* Rondani.

Mallophthiria Edwards, 1930:169. Type species: *Mallophthiria lanata* Edwards, 1930, by original designation. **Bombyliinae: Crocidiini.**

Malthacotricha Becker, 1907c:312. Type species: *Malthacotricha glauca* Becker, 1907, by monotypy. Synonymized with *Heterotropus* Loew, 1873, by Becker (1913a:482).

Mancia Coquillett, 1886b:159. Type species: *Mancia nana* Coquillett, 1886, by monotypy. **Anthracinae: Villini.**

Mandella Evenhuis, 1983a:207. Type species: *Systoechus albohirtus* Roberts, 1928, by original designation. **Bombyliinae: Dischistini.**

Mariobezzia Becker, 1913a:470. Type species: *Mariobezzia lichtwardti* Becker, 1913, by monotypy. **Cythereinae: Mariobezziini.**

Remarks. Previous works have treated the type fixation as *lichtwardti* by monotypy, thus implying that Becker (1913a) is prior to Becker (1913a). It is not quite as simple as that. Becker (1913b:503–654) describes 2 species (*lichtwardti* and *zarudnyi*) in *Mariobezzia* without giving a type-species designation. Since these 2 papers (Becker 1913a, b) were published simultaneously, the type fixation for this genus-group name depends on a first revisor determining which of these 2 publications and their corresponding nomenclatural acts was prior. Article 24 of the *Code* does not clarify whether one can consider a whole publication as a single nomenclatural act or whether one must take a narrow view and treat papers within a publication each as separate nomenclatural acts. I interpret the *Code* broadly and consider Becker (1913a) to be prior to Becker (1913b), thus the type of *Mariobezzia* is *lichtwardti* by monotypy.

Marleyimyia Hesse, 1956b:521. Type species: *Marleyimyia natalensis* Hesse, 1956, by original designation. **Anthracinae: Villoestrini.**

Marmasoma White, 1916:184, 188. Type species: *Marmasoma sumptuosa* White, 1916, by monotypy. **Ecliminae.**

Megacosmus† Cockerell, 1909a:10. Type species: *Megacosmus mirandus* Cockerell, 1909, by monotypy. Fossil shale genus. **Tomomyzinae.**

Meganedys Hull, 1973:322 (as subgenus of *Aphoebantus* Loew). Type species: *Aphoebantus petiolata* Hull, 1973, by original designation. Synonym of *Aphoebantus* Loew, 1872, **new synonymy.**

Remarks. Aphoebantus Loew is a considerably variable genus in the Nearctic Region. The characters of *Meganedys* Hull fall within the range of variation for *Aphoebantus*, and I can see no characters at present to support subgenera, thus the new synonymy here.

Meganthrax Evenhuis, 1979b:362. Type species: *Anthrax bipunctatus* Fabricius,

1805, by original designation. Synonymized with *Satyramoeba* Sack, 1909, by Evenhuis (1983c:324).

Megapalpus Macquart, 1834:394. Type species: *Phthiria capensis* Wiedemann, 1828, by monotypy. **Cythereinae: Mariobezziini.**

Remarks. See also remarks under *Dasypalpus* Macquart.

Megaphthiria Hall, 1976a:105 (as subgenus of *Phthiria* Meigen). Type species: *Phthiria pirioni* Edwards, 1937, by original designation. Raised to generic status and transferred from Phthiriinae to Bombyliinae by Evenhuis (1986:20). **Bombyliinae: Crocidiini.**

Melanderella† Cockerell, 1909c:70. Type species: *Melanderella glossalis* Cockerell, 1909, by monotypy. Fossil shale genus. **Systropodinae.**

Meomyia Evenhuis, 1983a:210. Type species: *Bombylius platyurus* Walker, 1849 [= *Bombylius fasciculatus* Macquart, 1840], by original designation. **Bombyliinae: Dischistini.**

Mesoclis Bezzi, 1921a:139 [1922a:105; 1922b:143; 1924:228, 229] (as subgenus of *Exoprosopa* Macquart). Type species: *Anthrax pygmalion* Fabricius, 1805, by original designation (as "varinerius-pygmalian, Fabr."). Treated as a subgenus of *Heteralonia* Rondani, 1863, by Bowden (1975b:318). **Anthracinae: Exoprosopini.**

Remarks. Though Bowden (1975b:318) treated this as a subgenus of *Heteralonia,* Zaitzev (1989:139) synonymized it with *Exoprosopa* Macquart, 1840. Bowden (1975b) gave convincing evidence to support *Mesoclis* as a subgenus of *Heteralonia,* which I follow here. Conventional dating of *Mesoclis* is from Bezzi (1924), however, the mention of the new genus-group name in combination with a valid species (*pygmalion* Fabricius) validates *Mesoclis* from Bezzi (1921a:139) by original designation. Bezzi (1924:229) gives *pygmalion* as the type by subsequent designation.

Metacosmus Coquillett, 1891:220. Type species: *Metacosmus exilis* Coquillett, 1891, by monotypy. **Tomomyzinae.**

Metapenta Bezzi, 1921a:138, 139 [1922c: 84; 1923:330; 1924:230] (as subgenus of *Exoprosopa* Macquart). Type species: *Exoprosopa pentala* Macquart, 1840, by subsequent designation of Bezzi (1924:230). Treated as a subgenus of *Heteralonia* Rondani, 1863, by Bowden (1975:318). **Anthracinae: Exoprosopini.**

Remarks. Bezzi (1921a) used 2 different spellings of this genus-group name (*Metapenta* on p. 138, 139 and *Metapeuta* on p. 2). I consider *Metapeuta* to be an incorrect original spelling of *Metapenta*. Though Bowden (1980:422) gives 1923 as the 1st valid publication of this genus, the 1921 work clearly satisfies the I.C.Z.N. criteria to make the genus-group name available.

Metapeuta Bezzi, 1921a:2. See *Metapenta* Bezzi.

Micomitra Bowden, 1964:147. Type species: *Exoprosopa parvicellula* Bezzi, 1921, by original designation. **Anthracinae: Exoprosopini.**

Micrusia Evenhuis, 1990a:38 (as subgenus of *Usia* Latreille). Type species: *Usia forcipata* Brullé, 1832, by original designation. **Usiinae: Usiini.**

Mima Meigen, 1820:175. Type species: *Anthrax phaeoptera* Meigen, 1820, by subsequent designation of Coquillett (1910:570). **Anthracinae: Exoprosopini.**

Remarks. Mima was originally proposed in synonymy but was made available by Coquillett's (1910:570) type designation and treatment of it as a senior synonym of *Exoprosopa* Macquart. Despite its being an available name, the type species of this genus could not be recognized until recently because of its vague original description. Publication of Meigen's color plates (Morge 1976) confirms the identity of *Anthrax phaeoptera* as very similar to, if not the same as, *Anthrax megerlei* Meigen, 1820, the type species of *Homolonia* Bowden, 1975 (currently a subgenus of *Heteralonia* Rondani). If the generic concept of *Mima* were to be accepted, it would be a senior synonym of *Homolonia* Bowden, 1975, and would be the nominate subgenus of the genus-group names currently within *Heteralonia* Rondani, 1863. An application to the I.C.Z.N. is in preparation to request suppression of *Mima* Meigen, 1820.

Mnemomyia Bowden, 1975b:308. Type species: *Mnemomyia rostrata* Bowden, 1975, by original designation. **Mythicomyiinae.**

Molybdamoeba Sack, 1909:519. Type species: *Anthrax tripunctata* Wiedemann *in* Meigen, 1820, by original designation. Synonymized with *Argyromoeba* Schiner, 1860, by Becker (1913a:448). Synonymized with *Spogostylum* Macquart, 1840, by Engel (1936a:438).

Mulio Latreille, 1796:155 (unnecessary new replacement name for *Cytherea* Fabricius; erroneously thought to be preoccupied by *Cythere* Mueller, 1785). Type species: *Cytherea obscura* Fabricius, 1794, by subsequent designation of Latreille

(1810:443). Junior synonymy under *Cytherea* Fabricius, 1794, was recognized by Fabricius (1805:29 [appendix]).

Remarks. Latreille's (1810) type-species designation was for *Mulio* Latreille, 1796, but due to the objective synonymy of the 2 genera, the designation is the same for *Cytherea,* the current name for this genus. See remarks under *Cytherea* Fabricius. Later type-species designations (all giving *obscura* Fabricius) include Blanchard *in* Audouin et al. (1845: pl. 168) and Becker (1913a:455).

Muscatheres Evenhuis, 1986:21. Type species: *Phthiria lurida* Walker, 1857, by original designation. **Bombyliinae: Dischistini.**

Myonema Roberts, 1929:563. Type species: *Myonema humile* Roberts, 1929, by original designation. **Tomomyzinae.**

Remarks. Evenhuis (1989:368) placed this genus in the tribe Myonematini of the Lomatiinae. Recent examination of specimens of this genus by me show characters that preclude it from being a member of the Lomatiinae. It is here transferred to the Tomomyzinae, with which it shares many more characters. This placement is only provisional until further study is done to clarify its proper place among the Bombyliidae.

Mythastrabes Hall and Evenhuis, 1986:340 (as subgenus of *Mythicomyia* Coquillett). Type species: *Mythicomyia antecessor* Melander, 1961, by original designation. **Mythicomyiinae.**

Mythenteles Hall and Evenhuis, 1986:332 (as subgenus of *Mythicomyia* Coquillett). Type species: *Mythicomyia mutabilis* Melander, 1961 [= *Empidideicus propleuralis* Melander, 1946, **new synonymy**], by original designation. Raised to generic status herein, **new status. Mythicomyiinae.**

Remarks. Originally proposed by Hall and Evenhuis (1986) as a subgenus of *Mythicomyia* Coquillett, 1893, subsequent research has elucidated characters that show *Mythenteles* to be a distinct genus and more closely related to *Empidideicus* Becker, 1907, than *Mythicomyia.* Raising *Mythenteles* to generic status results in the type species, *Mythenteles mutabilis* (Melander) and its senior synonym, *Mythenteles propleuralis* (Melander), becoming **new combinations.** See also remarks under *Cladella* Hall and Evenhuis.

Mythicomyia Coquillett, 1893:209. Type species: *Mythicomyia rileyi* Coquillett, 1893, by monotypy. **Mythicomyiinae.**

Remarks. Two spellings of this genus are found in the original publication (*Mythicomyia* and *Mythiocomyia,* both on p. 209). I consider *Mythicomyia* to be the correct original spelling of this genus.

With the removal herein of *Mythenteles* Hall and Evenhuis, 1986, from subgeneric status under *Mythicomyia,* there are now just 2 subgenera recognized in this genus: subg. **Mythicomyia** s. str. and subg. **Mythastrabes** Hall and Evenhuis, 1986.

Mythiocomyia. Coquillett, 1893:209. See *Mythicomyia.*

Neacreotrichus Cockerell, 1917:377. Type species: *Acreotrichus atratus* Coquillett, 1904 [= *Phthiria cingulata* Loew, 1846], by original designation. **Phthiriinae: Phthiriini.**

Remarks. This genus had long been considered a junior synonym of *Phthiria* Meigen, 1803. Evenhuis (1990a:47–49) resurrected *Neacreotrichus* from synonymy to accommodate Nearctic and Neotropical species previously misidentified as *Phthiria.*

Nectaropota Philippi, 1865:678. Type species: *Nectaropota setigera* Philippi, 1865, by monotypy. **Bombyliinae: Dischistini.**

Neobombylodes Evenhuis, 1978b:247 (new replacement name for *Bombylodes* Paramonov). Type species: *Dischistus multisetosus* Loew, 1857, automatic. Resurrected from synonymy with *Karakumia* Paramonov, 1927, by Zaitzev (1989:101). **Bombyliinae: Dischistini.**

Neodiplocampta Curran, 1934:193, 200. Type species: *Diplocampta roederi* Curran, 1931, by original designation. **Anthracinae: Villini.**

Remarks. According to Painter, Painter and Hall (1978:36–37), there are 2 subgenera recognized in this genus: subg. **Neodiplocampta** s. str. and subg. **Agitonia** Hull, 1966.

Neodischistus Painter, 1933b:1. Type species: *Neodischistus currani* Painter, 1933, by original designation. **Bombyliinae: Dischistini.**

Neosardus Roberts, 1929:560. Type species: *Neosardus principius* Roberts, 1929, by original designation. **Cythereinae: Mariobezziini.**

Remarks. This genus has been placed in the Cylleniinae by Hull (1973) and Evenhuis (1989). Examination of specimens of this genus by D. K. Yeates and myself shows that it is better placed in the Mariobezziini based on head and wing characters.

Nettadion Gistl, 1848:ix (unjustified new replacement name for *Lomatia* Meigen). Type species: *Anthrax belzebul* Fabricius, 1794, automatic. Synonymized with *Lomatia* Meigen, 1822, herein, **new synonymy.**

Neuria Newman, 1841:220. Type species: *Neuria lateralis* Newman, 1841, by subsequent designation of Hardy (1921:54). Junior homonym of *Neuria* Guenée, 1841. Synonymized with *Comptosia* Macquart, 1840, by Mik (1882:45).

Nevria Agassiz, 1847:249 (unjustified emendation of *Neuria* Newman). Type species: *Neuria lateralis* Newman, 1841, automatic. Synonymized with *Comptosia* Macquart, 1840, herein, **new synonymy.**

Nexus Hall and Evenhuis, 1987:623. Type species: *Nexus cyrtoidopsis* Hall and Evenhuis, 1987, by original designation. **Mythicomyiinae.**

Nomalonia Rondani, 1863:71 [1864:71]. Type species: *Cyllenia afra* Wiedemann, 1828 [misidentification, = *Nomalonia eremophila* Hesse, 1975], by original designation. **Cylleniinae: Enicini.**

Nothoschistus Bowden, 1985:103. Type species: *Bombylius transatlanticus* Philippi, 1865, by original designation. **Bombyliinae: Dischistini.**

Oestranthrax Bezzi, 1921a:130, 172 [1923:326]. Type species: *Anthrax obesa* Loew, 1863, by original designation. **Anthracinae: Villoestrini.**

Oestrimyza Hull, 1973:411. Type species: *Oestrimyza fenestrata* Hull, 1973, by original designation. **Anthracinae: Villoestrini.**

Ogcodera: Verrall *in* Scudder, 1882a:231. See *Ogcodocera* Macquart.

Ogcodocera Macquart, 1840:83 [1841:261]. Type species: *Ogcodocera dimidiata* Macquart, 1840 [= *Mulio leucoprocta* Wiedemann, 1828], by monotypy. **Lomatiinae: Lomatiini.**

Remarks. Scudder (1882a:231) lists *Ogcodera,* but this is merely a record of a subsequent misspelling of *Ogcodocera* and is not an emendation.

Ogilviella Paramonov, 1954:26. Type species: *Ogilviella tridentata* Paramonov, 1926, by monotypy. Synonymized with *Isotamia* Bezzi, 1912, by François

(1954:180). *Isotamia* treated as a subgenus of *Heteralonia* Rondani, 1863, by Bowden (1980:422).

Oligodranes Loew, 1844:160. Type species: *Oligodranes obscuripennis* Loew, 1844, by subsequent designation of Becker (1913a:484). **Oligodraninae.**

Remarks. This classification follows the most recent treatment of this genus by Evenhuis (1990a:66). African and North American species have previously been misidentified as *Oligodranes* and have been transferred to *Apolysis* Loew (see Evenhuis 1990a:32).

Onchopelma Hesse, 1938:973. Type species: *Onchopelma pulchella* Hesse, 1938, by original designation. **Mythicomyiinae.**

Oncodeocera Agassiz, 1847:257, 259 (unjustified emendation of *Ogcodocera* Macquart). Type species: *Ogcodocera dimidiata* Macquart, 1840 [= *Mulio leucoprocta* Wiedemann, 1828], automatic. Objective synonym of *Ogcodocera* Macquart, 1840.

Oncodocera Verrall *in* Scudder, 1882a:233 (unjustified emendation of *Ogcodocera* Macquart). Type species: *Ogcodocera dimidiata* Macquart, 1840 [= *Mulio leucoprocta* Wiedemann, 1828], automatic. Objective synonym of *Ogcodocera* Macquart, 1840.

Remarks. Oncodocera Bezzi (1924:3) is a later unjustified emendation of *Ogcodocera.*

Oncodosia Edwards, 1934:101. Type species: *Anthrax patula* Walker, 1849, by original designation. **Lomatiinae: Comptosiini.**

Oniromya Bezzi, 1921a:173. Incorrect original spelling. See *Oniromyia* Bezzi.

Oniromyia Bezzi, 1921a:71. Type species: *Eurycarenus pachyceratus* Bigot, 1892, by original designation. **Oniromyiinae.**

Remarks. The original publication gives 2 spellings for this genus (*Oniromyia* on p. 71; *Oniromya* on p. 173). I consider *Oniromya* to be an incorrect original spelling of *Oniromyia* Bezzi, 1921.

Opsonia Hull, 1973:306 (as subgenus of *Comptosia* Macquart). Type species: *Comptosia fenestrata* Hull, 1973, by original designation. Synonymized with *Comptosia* Macquart, 1840, by Evenhuis (1989b:367).

Oriellus Hull, 1973:403 (as subgenus of *Thyridanthrax* Osten Sacken). Type species: *Anthrax stigmulus* Klug, 1832, by original designation. Synonymized with *Exhyalanthrax* Becker, 1916, by Greathead (1981:317).

Ostentator Jaennicke, 1867:348. Type species: *Ostentator punctipennis* Jaennicke, 1867 [= *Bombylius sphenopterus* Loew, 1855], by monotypy. Synonymized with *Acrophthalmyda* Bigot, 1858, by Bigot (1892:336).

Othniomyia Hesse, 1938:707. Type species: *Othniomyia tylopelta* Hesse, 1938, by original designation. **Bombyliinae: Conophorini.**

Pachyanthrax François, 1964b:929. Type species: *Pachyanthrax villiersi* François, 1964, by original designation. **Anthracinae: Villini.**

Pachyneres Greene, 1924:62. Type species: *Pachyneres crassicornis* Greene, 1924, by monotypy. Synonymized with *Glabellula* Bezzi, 1902, implicitly by Bezzi (1926:256) and explicitly by Malloch (1928:607).

Pachysystropus† Cockerell, 1909b:56. Type species: *Pachysystropus rohweri* Cockerell, 1909, by monotypy. Fossil shale genus. ***Incertae sedis*** **(within Homoeophthalmae).**

Remarks. I have examined the type specimen of *Pachysystropus rohweri* Cockerell and find it very similar to *Amictus* Wiedemann, 1817, except that *Pachysystropus* has only 3 posterior cells versus 4 in *Amictus.* Correct placement will have to await further study of other specimens of *Pachysystropus.* It is very doubtful that *Pachysystropus* is a member of the Systropodinae as that subfamily is currently defined.

Palaeoamictus† Meunier, 1916:276. Type species: *Palaeoamictus spinosus* Meunier, 1916, by monotypy. Baltic amber genus. **Cylleniinae: Cylleniini.**

Palaeogeron† Meunier, 1914:195. Type species: *Palaeogeron vetustus* Meunier, 1914, by monotypy. Fossil shale genus. ***Incertae sedis*** **(within Homoeophthalmae).**

Remarks. The 4 posterior cells in the description and illustration of *Palaeogeron* precludes it from the Gerontinae. It appears closer to genera of Dischistini, but examination of the type specimen will be necessary before a correct generic placement can be made.

Palaeoplatypygus† Kovalev, 1985:186. Type species: *Palaeoplatypygus zaitzevi* Kovalev, 1985, by original designation. Fossil chalk genus. **Mythicomyiinae.**

Remarks. Kovalev used 2 different spellings in the original publication of this genus (*Palaeoplatypygus* on p. 186, 187; *Paleoplatypygus* on p. 186). I consider *Paleoplatypygus* to be an incorrect original spelling of *Palaeoplatypygus* Kovalev, 1985. This is the oldest known fossil Bombyliidae, dating from the Siberian Jurassic.

Paleoplatypygus: Kovalev, 1985:186. Incorrect original spelling. See *Palaeoplatypygus* Kovalev.

Palintonus François, 1964a:323. Type species: *Lepidophora austeni* Paramonov, 1949, by original designation. **Ecliminae.**

Pantarbes Osten Sacken, 1877:226, 229, 254. Type species: *Pantarbes capito* Osten Sacken, 1877, by monotypy. **Cythereinae: Cythereini.**

Pantostomus Bezzi, 1921b:474 [1922c:79]. Type species: *Pantostomus gibbiventris* Bezzi, 1921, by monotypy. **Tomomyzinae.**

Parabombylius Williston, 1907:1. Type species: *Thlipsogaster ater* Coquillett, 1894 [preoccupied, = *Bombylius coquilletti* Williston, 1901], by subsequent designation of Coquillett (1910:583). Treated as a subgenus of *Bombylius* Linnaeus, 1758, by Hall and Evenhuis (in press). **Bombyliinae: Bombyliini.**

Parachistus Greathead, 1980:238. Type species: *Parachistus pulchellus* Greathead, 1980, by original designation. **Bombyliinae: Dischistini.**

Paraconsors Hall and Evenhuis, 1987:606. Type species: *Mythicomyia aperta* Melander, 1961, by original designation. **Mythicomyiinae.**

Remarks. According to Hall and Evenhuis (1987), there are currently 2 subgenera recognized in this genus: subg. **Paraconsors** s. str. and subg. **Elachymyia** Hall and Evenhuis, 1987.

Paracorsomyza† Hennig, 1966:4. Type species: *Corsomyza crassirostris* Loew, 1850 (as "*Corsomyia*"), by monotypy. Baltic amber genus. **Cythereinae: Mariobezziini.**

Paracosmus Osten Sacken, 1877:262 (new replacement name for *Allocotus* Loew). Type species: *Allocotus edwardsii* Loew, 1872, automatic. **Tomomyzinae.**

Remarks. There are 2 subgenera currently considered in this genus based on Hull (1973): subg. **Paracosmus** s. str. and subg. **Actherosia** Hull, 1973.

Paradiplocampta Hall, 1975a:110. Type species: *Paradiplocampta tabeti* Hall, 1975, original designation. **Anthracinae: Villini.**

Paradosia Hull, 1973:306 (as subgenus of *Comptosia*). Type species: *Comptosia duofasciata* Hull, 1973, by original designation. Raised to generic status and synonymized with *Comptosia* Macquart, 1840, by Evenhuis (1989b:367).

Parageron Paramonov, 1929:189. Type species: *Parageron orientalis* Paramonov, 1929 [= *Usia lutescens* Bezzi, 1925], by monotypy. **Usiinae: Usiini.**

Paramonovella: Hull, 1973:61. Type species: *Bombylius platyurus* Walker, 1849 [= *Bombylius fasciculatus* Macquart, 1840], by present designation of 1st species included herein. *Nomen nudum.* Objective synonym of *Meomyia* Evenhuis, 1983.

Remarks. Examination of bombyliids in the Australian National Insect Collection (ANIC), Canberra, and the Canadian National Collection, Ottawa, revealed some specimens with handwritten labels stating "Paramonovella" that have been found to be congeneric with *Meomyia* Evenhuis, 1983. *Paramonovella* Hull does not satisfy the criteria for availability and is a *nomen nudum.* The name is only mentioned in Hull's (1973:61) list of genera, thus *Meomyia* Evenhuis, 1983, is the valid name for species in this genus.

Paramonovella: Zaitzev, 1982a:521 [1982b:68]. Type species: *Bombylius albiceps* Macquart, 1848 (as *Paramonovella albiceps*), by monotypy. *Nomen nudum.* Synonym of *Meomyia* Evenhuis, 1983.

Remarks. This genus was listed among genera and species surveyed by Zaitzev for his paper on the closing structure of the pseudotrachea of flies. He obviously did not intend to describe this genus (he undoubtedly had gotten the genus name, *Paramonovella,* from label data of specimens in the ANIC when he visited there in 1979).

Paramonovomyia Evenhuis, 1980:329. Type species: *Paramonovomyia halli* Evenhuis, 1980, by original designation. Synonymized with *Aleucosia* Edwards, 1934, by Yeates (in litt.).

Remarks. See also remarks under *Chelina* Hull.

Paranthracina Paramonov, 1933:56 [1934a:27; 1936:31] (new replacement name for *Paranthrax* Paramonov). Type species: *Paranthrax africanus* Paramonov, 1931 [= *Hyperalonia niveifrons* Bezzi, 1914], automatic. Synonymized with *Ligyra* Newman, 1841, by Bowden (1971:5).

Paranthrax Bigot, 1876a:71 (20 April) [1876b:lxvi (11 October)]. Type species: *Anthrax rufiventris* Blanchard, 1852, by original designation. **Anthracinae: Villini.**

Remarks. This genus has been traditionally dated from the pages numbered with Roman numerals in the *Bull. Soc. Entomol. Fr.* An earlier version of the *Bulletin* (abbreviated in the Literature Cited section as *Bull. Bimens. Soc. Entomol. Fr.*) has the 1st included valid publication for the genus. There is no change in spelling or type species.

Paranthrax Paramonov, 1931:57. Type species: *Paranthrax africanus* Paramonov, 1931 [= *Hyperalonia niveifrons* Bezzi, 1914], by monotypy. Junior homonym of *Paranthrax* Bigot, 1876. Synonymized with *Ligyra* Newman, 1841, by Bowden (1971:5).

Parasystoechus Hall, 1976a:27 (as subgenus of *Bombylius*). Type species: *Bombylius flavescens* Hall, 1976, by original designation. Raised to generic status by Zaitzev (1984a:35 and fig. 11; 1984b:171 and pl. 1, fig. 3). **Bombyliinae: Dischistini.**

Remarks. Zaitzev (1984a, b) raised this subgenus to generic status without explanation and included an Australian species (*Bombylius australis* Guérin-Méneville, 1831) belonging to a different, but as yet undescribed, genus of Dischistini (see Evenhuis 1989b:365). I assume that Zaitzev's (1984a, b) *Parasystoechus* is a misidentification of Hall's (1976) genus and not an inadvertent validation of a new genus (see, for example, *Paramonovella* Zaitzev). Bowden (1985:103) more explicitly raised *Parasystoechus* Hall, 1976, to generic status and provided evidence for his action.

Paratoxophora Engel, 1936b:39. Type species: *Paratoxophora cuthbertsoni* Engel, 1936, by monotypy. **Bombyliinae: Dischistini.**

Paravilla Painter, 1933a:10 (as subgenus of *Villa* Lioy). Type species: *Villa edititoides* Painter, 1933, by original designation. Raised to generic status by Hull (1973:365) and again by Hall (1981:2). **Anthracinae: Exoprosopini.**

Parisus Walker, 1852:196. Type species: *Parisus paterculus* Walker, 1852, by monotypy. Synonymized with *Bombylius* Linnaeus, 1758, by Loew (1860b:248).

Peringueyimyia Bigot, 1886b:cx (7 August) [1886d:cx (15 October)]. Type species: *Peringueyimyia capensis* Bigot, 1886, by monotypy. **Cylleniinae: Peringueyimyiini.**

Petrorossia Bezzi, 1908:35. Type species: *Bibio hesperus* Rossi, 1790, by original designation. **Anthracinae: Petrorossiini.**

Phthiria Meigen, 1803:268. Type species: *Bombylius pulicarius* Mikan, 1796, by monotypy. **Phthiriinae: Phthiriini.**

Remarks. Scudder (1882a), based on a list of names supplied to him by Verrall, listed 4 different names (*Phthyria, Phtiria,* and *Phtyria,* on p. 261; *Pthiria* on p. 286), which he synonymized with *Phthiria.* These are all merely listings of subsequent misspellings of *Phthiria* and are not considered here as emendations.

Phthyria: Verrall *in* Scudder, 1882a:261. See *Phthiria* Meigen.

Phtiria: Verrall *in* Scudder, 1882a:261. See *Phthiria* Meigen.

Phtyria: Verrall *in* Scudder, 1882a:261. See *Phthiria* Meigen.

Pilosia Hull, 1973:137. Type species: *Pilosia flavipilosa* Hull, 1973, by original designation. **Bombyliinae: Dischistini.**

Pioperna Enderlein, 1926:70. Type species: *Cephenus femoratus* Karsch, 1880, by original designation. Reduced to subgeneric status by Hull (1973:244). Synonym of the subgenus *Systropus (Systropus)* Wiedemann, 1820, **new synonymy.**

Remarks. Pioperna Enderlein seems to have been ignored by recent catalogers and revisors of the genus *Systropus.* Carrera and d'Andretta (1950) did not refer to it nor to Enderlein (1926) in their revision of Brazilian species; Painter and Painter (1963) did not mention it in their revision of North American species; Bowden (1967) did not mention it in his Afrotropical revision; and Painter, Painter and Hall (1978) omitted it from their catalog of Neotropical Bombyliidae. Hull (1973:244) listed *Pioperna* as a subgenus under *Systropus* but gave no further details. Painter and Painter (1974:74) inadvertently mentioned it in citing the label data of the type specimen of *femoratus.* Apparently not realizing they were dealing with a generic name, they cited the label data as "properna *femorata* ♂ Type (Karsch) 1880, Dr. Enderlein." *Pioperna* was erected for *femoratus,* which does not differ from other species of the subgenus *Systropus* enough to warrant separate generic or subgeneric status, thus the new synonymy here.

Pipunculopsis Bezzi, 1925:210, 211. Type species: *Pipunculopsis bivittata* Bezzi, 1925, by original designation. **Anthracinae: Petrorossiini.**

Platamodes Loew, 1855:40 (as subgenus of *Bombylius*). Type species: *Bombylius depressus* Loew, 1855, by monotypy. Raised to generic status by Bigot (1858:578). Junior homonym of *Platamodes* Ménétriés, 1849. Objective synonym of *Platamomyia* Brèthes, 1926.

Platamomyia Brèthes, 1926:105 (new replacement name for *Platamodes* Loew). Type species: *Bombylius depressus* Loew, 1855, automatic. **Bombyliinae: Conophorini.**

Remarks. Hall (1975) treats this genus as related to *Conophorus* Meigen, but placed it in a broad concept of Bombyliinae. It is treated here in Conophorini.

Platygaster Zetterstedt, 1838:574. Type species: *Platygaster arcticus* Zetterstedt, 1838, by monotypy. Junior homonym of *Platygaster* Latreille, 1809. Objective synonym of *Glabellula* Bezzi, 1902.

Remarks. Platygaster Zetterstedt, 1838, placed on the Official Index of Rejected and Invalid Generic Names in Zoology by I.C.Z.N. (1989:148).

Platypygus Loew, 1844:127. Type species: *Platypygus chrysanthemi* Loew, 1844, by monotypy. **Mythicomyiinae.**

Plesiocera Macquart, 1840:82 [1841:360]. Type species: *Plesiocera algira* Macquart, 1840, by monotypy. **Anthracinae: Plesiocerini.**

Plichtamyia Evenhuis, 1979a:39. Type species: *Geron dispar* Macquart, 1850, by original designation. Subgenus of *Geron* Meigen, 1820. **Gerontinae.**

Ploas Latreille, 1804:190. Type species: *Ploas hirticornis* Latreille, 1805 [= *Bombylius virescens* Fabricius, 1787], by subsequent monotypy in Latreille (1805:300). Synonymized with *Conophorus* Meigen, 1803, by Kertész (1909:103).

Remarks. Fabricius (1805:136) includes 2 species in *Ploas: Bombylius virescens* Fabricius and *Bombylius maurus* Mikan (as "*Conophorus maurus*") in the same year as Latreille (1805:300). Fabricius's (1805) work has not been accurately dated within the year 1805, but the preface was dated 10 March. Latreille's (1805) work has been dated between 20 February and 21 March by Dupuis (1986). If Fabricius's (1805) work was to be accepted as prior to Latreille (1805), a subsequent type-species designation from the 1st included 2 species included by Fabricius (1805:136) would be necessary. It is highly unlikely that Fabricius's work could have come out before 21 March 1805—only 11 days after the date of the preface—hence, until future dating of these publications prove otherwise, I consider Latreille's (1805) work to predate Fabricius (1805).

Podolepida Hull, 1973:386 (as subgenus of *Lepidanthrax* Osten Sacken). Type species: *Lepidanthrax lutzi* Curran, 1930, by original designation. Raised to generic status and synonymized with *Lepidanthrax* Osten Sacken, 1886, by Hall (1976b:290).

Poecilanthrax Osten Sacken, 1886b:112 (as subgenus of *Anthrax* Scopoli), 119 (as "group" of *Anthrax* Scopoli)]. Raised to generic status by Painter (1930:803). Type species: *Anthrax alcyon* Say, 1824, by subsequent designation of Coquillett (1910:593). **Anthracinae: Villini.**

Remarks. Most listings of *Poecilanthrax* (e.g., Painter & Hall 1960; Painter & Painter 1965) state it was originally described as a "group" of *Anthrax* (as it is in Osten Sacken 1886b:119). However, Osten Sacken (1886b:112) specifically states that it is a subgenus.

Poecilognathus Jaennicke, 1867:350. Type species: *Poecilognathus thlipsomyzoides* Jaennicke, 1867, by monotypy. **Phthiriinae: Poecilognathini.**

Popsia Costa, 1863:51. Type species: *Popsia ridibunda* Costa, 1863, by monotypy. Synonymized with *Platypygus* Loew, 1844, by Loew (1873:206).

Praecytherea† Théobald, 1937:168. Type species: *Praecytherea sardii* Théobald, 1937, by monotypy. Fossil shale genus. **Nomen dubium. Cythereinae: *Tribus inquirendus.***

Remarks. I have not seen the type specimen and the original description is rather vague. Until it is examined, it must remain a *nomen dubium* and its tribal classification within the Cythereinae is *tribus inquirendus*.

Procyrtosia† Zaitzev, 1986:817 [1987:152]. Type species: *Procyrtosia sukatshevae* Zaitzev, 1986, by original designation. Siberian amber genus. **Mythicomyiinae.**

Proglabellula† Hennig, 1966:15. Type species: *Proglabellula electrica* Hennig, 1966, by monotypy. Baltic amber genus. Synonymized with *Glabellula* Bezzi, 1902, herein, **new synonymy.**[1]

1. Since preparing this paper, I have had discussions with H. Schumann, Humboldt-Universität, Berlin, who has a paper in press in which he also proposes the synonymy of *Proglabellula* with *Glabellula.*

Remarks. I have examined the type of *Proglabellula electrica* Hennig, 1966, and the 2nd specimen of *electra* referred to in Hennig (1967:13) and can find no characters that distinguish this genus from the extant *Glabellula* Bezzi, 1902, thus the new synonymy.

Proplatypygus† Hennig, 1969:58. Type species: *Proplatypygus succineus* Hennig, 1969, by monotypy. Baltic amber genus. **Mythicomyiinae.**

Remarks. The Baltic amber type specimen is dated from the Tertiary. Zaitzev (1986:815) adds a 2nd species of this genus from Siberian Cretaceous amber.

Prorachthes Loew, 1869a:381. Type species: *Prorachthes ledereri* Loew, 1869, by monotypy. **Bombyliinae: Conophorini.**

Prorates Melander, 1906:372. Type species: *Prorates claripennis* Melander, 1906, by monotypy. **Proratinae.**

Remarks. See remarks under *Alloxytropus* Bezzi and *Caenotoides* Hall.

Prorostoma Hesse, 1956a:121. Type species: *Plesiocera integra* Bezzi, 1922, by original designation. **Anthracinae: Plesiocerini.**

Protepacmus† Cockerell, 1916:94. Type species: *Protepacmus setosus* Cockerell, 1916, by monotypy. Fossil shale genus. Synonym of *Villa* Lioy, 1864, **new synonymy.**

Remarks. I have examined the type specimen of *Protepacmus setosus* and can find no differences separating it from typical *Villa,* thus the new synonymy.

Prothaplocnemis Bezzi, 1925:218, 220. Type species: *Argyromoeba anthracina* Becker, 1902 (as "*Argyramoeba*"), by original designation. **Anthracinae: Petrorossiini.**

Protolomatia† Cockerell, 1914b:723. Type species: *Protolomatia antiqua* Cockerell, 1914, by monotypy. Fossil shale genus. Synonym of *Paracosmus* Osten Sacken, 1877, **new synonymy.**

Remarks. I have examined the type specimens of *Protolomatia antiqua* Cockerell, 1914, and *P. recurrens* Cockerell, 1916, and can find no characters that distinguish these 2 species from species of *Paracosmus,* thus the new synonymy.

Protophthiria† Cockerell, 1914b:720. Type species: *Protophthiria palpalis* Cockerell, 1914, by monotypy. Synonymized with *Lithocosmus* Cockerell, 1909, by Evenhuis (1984:157).

Psamatamoeba Sack, 1909:536. Type species: *Anthrax isis* Meigen, 1820, by original designation. Synonymized with *Argyromoeba* Schiner, 1860, by Becker (1913a:448). Synonymized with *Spogostylum* Macquart, 1840, by Engel (1936a:438).

Pseudempis Bezzi, 1921a:94. Type species: *Amictus heteropterus* Wiedemann, 1821 [misidentification, = *Pseudoamictus bezzii* Paramonov, 1930], by original designation and with *Pseudoammictus* Bigot, 1892, by Bezzi (1924:20).

Pseudoammictus Bigot, 1892:342. Type species: *Amictus heteropterus* Wiedemann, 1821 (as *Ammictus*), by monotypy. Reduced to subgeneric status under *Geron* Meigen, 1820, by Bowden (1974:91). **Gerontinae.**

Remarks. Pseudoamictus is a common subsequent misspelling of *Pseudoammictus* and is not an emendation.

Pseudogeron Cresson, 1915:201. Type species: *Pseudogeron mitis* Cresson, 1915, by original designation. Synonymized with *Apolysis* Loew, 1860, by Evenhuis (1990a:27).

Pseudoglabellula Hesse, 1967:118. Type species: *Pseudoglabellula meridionalis* Hesse, 1967, by original designation. **Mythicomyiinae.**

Pseudopenthes Roberts, 1928:132. Type species: *Pseudopenthes fenestrata* Roberts, 1928, by original designation. **Anthracinae: Villini.**

Psiatholasius Becker, 1906:145. Type species: *Psiatholasius bombyliiformis* Becker, 1906, by original designation. Synonymized with *Legnotus* Loew, 1855, by Becker (1907b:454). Synonymized with *Legnotomyia* Bezzi, 1902, by Paramonov (1931:116).

Psiloderoides Hesse, 1967:121. Type species: *Psiloderoides mansfieldi* Hesse, 1967, by original designation. **Mythicomyiinae.**

Pteraulacodes Hesse, 1956a:348. Type species: *Pteraulacodes karooensis* Hesse, 1956, by original designation. **Anthracinae: Aphoebantini.**

Pteraulax Bezzi, 1921a:117. Type species: *Pteraulax flexicornis* Bezzi, 1921, by original designation. **Anthracinae: Aphoebantini.**

Pterobates Bezzi, 1921a:130 [1922c:85; 1924:273] (as subgenus of *Exoprosopa* Macquart). Type species: *Anthrax apicalis* Wiedemann, 1821, by monotypy. Raised to generic status by Bowden (1980:425). **Anthracinae: Exoprosopini.**

Remarks. Bezzi proposed the name *Pterobates* 3 times (1921a:130; 1922c:85; 1924:273). The 1st and 2nd proposals included only 1 species (*Anthrax apicalis* Wiedemann) and a brief diagnosis. The 3rd included 3 species (*apicalis, pennipes* Wiedemann, and *chalybea* Röder) with *apicalis* designated as the type species.

Ptimia Rafinesque, 1815:221 (unjustified emendation of *Phthiria* Meigen). Type species: *Bombylius pulicarius* Mikan, 1796, automatic. Synonym of *Phthiria* Meigen, 1803, **new synonymy.**

Pthiria: Verrall *in* Scudder, 1882a:286. See *Phthiria* Meigen.

Pusilla Paramonov, 1954:27. Type species: *Pusilla longirostris* Paramonov, 1954, by original designation. **Cythereinae: Mariobezziini.**

Pygocona Hull, 1973:195 (as subgenus of *Phthiria* Meigen). Type species: *Pygocona flavicincta* Hull, 1973 [= *Phthiria flava* Hardy, 1933], by original designation. Raised to generic status by Evenhuis (1986:15). **Phthiriinae: Phthiriini.**

Relictiphthiria Evenhuis, 1986:17. Type species: *Relictiphthiria primogenita* Evenhuis, 1986, by original designation. **Phthiriinae: Poecilognathini.**

Rhabdopselaphus Bigot, 1886a:ciii (3 July) [1886b:ciii (15 October)]. Type species: *Rhabdopselaphus mus* Bigot, 1886, by monotypy. Synonymized with *Apolysis* Loew, 1860, by Evenhuis (1990a:27).

Rhynchanthrax Painter, 1933a:6 (as subgenus of *Villa* Lioy). Type species: *Anthrax parvicornis* Loew, 1869, by original designation. Raised to generic status by Hull (1973:365) and again by Hall (1981:2). **Anthracinae: Villini.**

Robertsella Evenhuis, 1978b:247 (new replacement name for *Cyrtomorpha* White). Type species: *Cyrtomorpha paganica* White, 1916, automatic. Junior homonym of *Robertsella* Guinot, 1969. Objective synonym of *Acridophagus* Evenhuis, 1983.

Satyramoeba Sack, 1909:517. Type species: *Anthrax etrusca* Fabricius, 1794, by original designation. **Anthracinae: Anthracini.**

Scinax Loew, 1855:42 (as subgenus of *Bombylius* Linnaeus). Type species: *Bombylius sphenopterus* Loew, 1855, by original designation. Raised to generic status by Bigot (1858:584). Junior homonym of *Scinax* Wagler, 1830. Synonymized with *Acrophthalmyda* Bigot, 1858, by Bigot (1859:209).

Semiramis Becker, 1913a:586. Type species: *Semiramis punctipennis* Becker, 1913, by original designation. **Bombyliinae: Crocidiini.**

Semistoechus Hall, 1976a:27 (as subgenus of *Bombylius* Linnaeus). Type species: *Bombylius ovatus* Hall, 1976, by original designation. Raised to generic status by Evenhuis (1991:165).

Remarks. Two different spellings for this genus-group name are found in the original publication (*Semistoechus* on p. 17, 265; *Semistochus* on p. 26–28, 31). I consider *Semistochus* to be an incorrect original spelling of *Semistoechus* Hall, 1976.

Semistochus: Hall, 1976a:26. Incorrect original spelling. See *Semistoechus* Hall.

Sericosoma Macquart, 1850:419(115). Type species: *Sericosoma fascifrons* Macquart, 1850, by original designation. **Cythereinae: Cythereini.**

Sericothrix Hall, 1976a:59. Type species: *Sericothrix dioedes* Hall, 1976, by original designation. **Cythereinae: Cythereini.**

Remarks. Hall used 2 different spellings for this genus (*Bombylthrix* on p. 10; *Sericothrix* on p. 59). *Bombylthrix* was an inadvertent error for *Sericothrix* (Hall, in litt.) and is treated here as an incorrect original spelling.

Sericusia Edwards, 1937:36. Type species: *Sericusia lanata* Edwards, 1937, by original designation. **Bombyliinae: Dischistini.**

Sesyomyia: Bezzi, 1921a:173. Incorrect original spelling. See *Sosiomyia* Bezzi.

Sinaia Becker, 1916:65. Type species: *Sinaia kneuckeri* Becker, 1916, by original designation. **Cylleniinae: Cylleniini.**

Remarks. Though Zaitzev (1989) places this genus in the Ecliminae, neither Hall

(1969) nor Greathead (1988) mention *Sinaia* in their reclassifications of Ecliminae. I follow Hull (1973) in his placement of *Sinaia* within the Cylleniini.

Sinaia: Hermann, 1909:73. Type species: *Sinaia kneuckeri* Becker, 1916, by present designation of 1st species included herein. *Nomen nudum.* Synonym of *Sinaia* Becker, 1916.

Remarks. Though Hermann (1909:73) lists this name and the only species included *(kneuckeri)* as "n. g. et sp." there is not enough information presented to make either the genus name or the species name available in this publication.

Sisyromyia White, 1916:192, 197. Type species: *Bombylius auratus* Walker, 1849, by subsequent designation of Bezzi (1924:7). **Bombyliinae: Dischistini.**

Sisyrophanus Karsch, 1886:53. Type species: *Sisyrophanus homeyeri* Karsch, 1886, by monotypy. **Bombyliinae: Dischistini.**

Sobarus Loew, 1855:6, 39 (as subgenus of *Bombylius* Linnaeus). Type species: *Bombylius anomalus* Wiedemann, 1821, by original designation. Raised to generic status by Bigot (1858:578). Synonymized with *Adelidea* Macquart, 1840, by Schiner (1868:114).

Sosiomya: Bezzi, 1921a:180. Incorrect original spelling. See *Sosiomyia* Bezzi.

Sosiomyia Bezzi, 1921a:67. Type species: *Sosiomyia comata* Bezzi, 1921, by original designation (as *"carnata"*). **Bombyliinae: Conophorini.**

Remarks. Bezzi used 3 different spellings for the genus in the original publication (*Sosiomyia* on p. 67; *Sesyomyia* on p. 173; *Sosiomya* on p. 180—explanation of plates). I consider *Sesyomyia* and *Sosiomya* to be incorrect original spellings of *Sosiomyia.*

Bezzi also used 2 different spellings for his type species of *Sosiomyia* (*carnata* on p. 69, 173, 176; *comata* on p. 180—the explanation of the plates). Bezzi (1921b:470; 1924:19) listed *comata* as the only species included in *Sosiomyia.* Bezzi, in his errata (1921b:477), revised the 2 spellings by correcting *carnata* to *comata.* Hesse (1938:703) noted the 2 different spellings of this species but cited *comata* as an incorrect subsequent spelling in Bezzi (1921b:470), not an original spelling, and continued to use *carnata* as the spelling for this species, an incorrect usage that was followed by others.

Bowden (1980:401) treats this genus in the Dischistini. Hall (1975) compares it to *Conophorus* Meigen and *Platamomyia* Brèthes, with which I concur. It is thus placed here in the Conophorini.

Sparnopolius Loew, 1855:43 (as subgenus of *Bombylius* Linnaeus). Type species: *Bombylius fulvus* Wiedemann, 1821 [preoccupied, = *Bombylius confusus* Wiedemann, 1824], by subsequent designation of Coquillett (1910:606). Raised to generic status by Bigot (1858:584). **Bombyliinae: Conophorini.**

Sphaerogaster Zetterstedt, 1842:232 (new replacement name for *Platygaster* Zetterstedt). Type species: *Platygaster arcticus* Zetterstedt, 1838, automatic. Junior homonym of *Sphaerogaster* Sturm, 1826. Objective synonym of *Glabellula* Bezzi, 1902.

Remarks. Sphaerogaster Zetterstedt, 1842, placed on the Official Index of Rejected and Invalid Generic Names in Zoology by I.C.Z.N. (1989:148).

Sphenoidoptera Williston, 1901:295. Type species: *Sphenoidoptera varipennis* Williston, 1901, by monotypy. **Cylleniinae: Cylleniini.**

Remarks. Painter, Painter and Hall (1978:23) synonymized this genus under *Cyllenia* Latreille, 1802. I don't believe that *Cyllenia* occurs in the New World and follow Theodor (1983:139) in retaining *Sphenoidoptera* as a distinct genus.

Spogastylum: Macquart, 1840: pl. 19. Incorrect original spelling. See *Spogostylum* Macquart.

Spogostylum Macquart, 1840:53 [1841:331]. Type species: *Spogostylum mystaceum* Macquart, 1840 [= *Anthrax tripunctata* Wiedemann in Meigen, 1820], by monotypy. **Anthracinae: Anthracini.**

Remarks. Macquart used 2 different spellings for this genus in the original publication (*Spogostylum* on p. 53; *Spogastylum* on pl. 19). I consider *Spogastylum* to be an incorrect original spelling of *Spogostylum.* For clarification of the type-species identity and resulting generic concept of this genus, see Evenhuis (1985a).

Spongostylum Agassiz, 1847:349 (unjustified emendation of *Spogostylum* Macquart). Type species: *Spogostylum mystaceum* Macquart, 1840 [= *Anthrax tripunctata* Wiedemann *in* Meigen, 1820], automatic. Objective synonym of *Spogostylum* Macquart, 1840.

Staurostichus Hull, 1973:76, 104. Type species: *Staurostichus chiastoneura* Hull, 1973, by original designation. **Bombyliinae: Dischistini.**

Remarks. Evenhuis (1983a) designated a neotype for *Staurostichus* after giving evidence of its possible loss and destruction. A personal search was conducted in December 1989 of the Hull Collection in the CNC for the possible existence of "lost" Hull types. The search turned up most of the previously thought "lost" types including the type specimens of *Staurostichus chiastoneura* Hull, 1973. A single male and female were without type labels but had pencilled labels with *"Staurostichus"* in Hull's hand and were collected from the type locality mentioned in Hull (1973:104–05). An application to the I.C.Z.N. is required to rule on whether or not the neotype is to be retained [Art. 75(h)].

Stenolonia Bowden, 1975b:319 (as subgenus of *Heteralonia* Rondani). Type species: *Exoprosopa loxospila* Hesse, 1956, by original designation. Synonymized with the subgenus *Heteralonia (Isotamia)* Bezzi, 1912, by Bowden (1980:422).

Stomylomyia Bigot, 1887a:xxxi (19 February) [1887b:xxxi (25 May)]. Type species: *Stomylomyia leonina* Bigot, 1887 [= *Tomomyza europaea* Loew, 1869], by monotypy. **Anthracinae: Plesiocerini.**

Remarks. See Macgillavry (1936) and Evenhuis et al. (1989) for explanation of the different issues of the *Bull. Bimens. Soc. Entomol. Fr.* and the *Bull. Soc. Entomol. Fr.* Smith et al. (1980) dated this article as 1888, but a check of mailing dates of each version of this issue of the *Bulletin* shows that they were both issued in 1887.

Stonyx Osten Sacken, 1886a:80, 94. Type species: *Stonyx clotho* Osten Sacken, 1886, by subsequent designation of Coquillett (1910:609). **Anthracinae: Villini.**

Stygia Meigen, 1820:137. Type species: *Anthrax belzebul* Fabricius, 1794, by subsequent designation of Westwood (1840:131) [type designation for *Lomatia* Meigen, 1822, with *Stygia* in synonymy]. Junior homonym of *Stygia* Latreille, 1804. Objective synonym of *Lomatia* Meigen, 1822.

Stygide: Latreille, 1825:491 (new replacement name for *Stygia* Meigen). Unavailable name. Vernacular name without status in nomenclature.

Stygides Latreille, 1829:465 (new replacement name for *Stygia* Meigen). Type species: *Anthrax belzebul* Fabricius, 1794, automatic. Synonymized herein with Lomatia Meigen, 1822, **new synonymy.**

Remarks. Guérin-Méneville (1829:692) mentioned the name *Stygides* in the same year, attributing the name to Latreille. Dating of the 2 works shows that Latreille's proposal of this name (11 April 1829) preceded Guérin-Méneville's (May 1829) by just a few weeks.

Stygidium Berthold, 1827:500 (new replacement name for *Stygia* Meigen). Type species: *Anthrax belzebul* Fabricius, 1794, automatic. Synonymized with *Lomatia* Meigen, 1822, herein, **new synonymy.**

Remarks. Sherborn (1931:6213) and Neave (1940b:339) incorrectly attribute this name to the Coleoptera.

Stymphalina: Hull, 1973:445. Type species: *Nemotelus tigrinus* De Geer, 1776 (as "*Anthrax tigrinus* Fabricius"), by monotypy. *Nomen nudum.* Synonym of *Xenox* Evenhuis, 1985.

Remarks. Though Hull included a single described species (*Nemotelus tigrinus* De Geer, 1776) in *Stymphalina* and remarked that the ecology and behavior (viz., the host of parasitization) are similar to *Walkeromyia,* there is not enough information in the original proposal of Hull's *Stymphalina* to validate it as a genus-group name.

Subbombylius Kapoor *in* Kapoor, Agarwal and Grewal, 1979:411. Type species: *Subbombylius bowdeni* Kapoor and Agarwal, 1979, by original designation. Synonym of *Amictus* Wiedemann, 1817, **new synonymy.**

Remarks. Two different spellings were used in the original publication of this genus (*Subbombylius* on p. 411, 416; *Sybbombylius* on p. 411). I consider *Sybbombylius* to be an incorrect original spelling of *Subbombylius.* I agree with Dr. David Greathead (in litt.) that the description and illustration of *Subbombylius* give convincing evidence that it is a junior synonym of *Amictus* Wiedemann, 1817, thus the new synonymy.

Sybbombylius: Kapoor *in* Kapoor, Agarwal and Grewal, 1979:411. Incorrect original spelling. See *Subbombylius* Kapoor.

Symballa Enderlein, 1926:70 (as genus). Type species: *Systropus leptogaster* Loew, 1860 [misidentification, = *Systropus holaspis* Speiser, 1914], by original designation. Reduced to subgeneric status and synonymized with the subgenus *Systropus (Systropus)* Wiedemann, 1820, by Bowden (1967:144).

Synthesia Bezzi, 1921a:130. Type species: *Synthesia fucoides* Bezzi, 1921, by original designation. **Anthracinae: ?Villini.**

Syrphoides Evenhuis, 1983a:206. Type species: *Bombylius byrrhus* Bowden, 1971, by original designation. Junior homonym of *Syrphoides* Hippa, 1968. Synonym of *Eristalopsis* Evenhuis, 1985.

Systoechus Loew, 1855:34 (as subgenus of *Bombylius* Linnaeus). Type species: *Bombylius sulphureus* Mikan, 1796, by subsequent designation of Coquillett (1910:611). Raised to generic status by Bigot (1858:584). **Bombyliinae: Bombyliini.**

Systrophopus Karsch, 1880:656 (unjustified emendation of *Systropus* Wiedemann). Type species: *Systropus macilentus* Wiedemann, 1820, automatic. See *Systropus* Wiedemann, 1820.

Systropus Henning, 1832:335. Type species: *Systropus macilentus* Henning, 1832 [preoccupied, = *Systropus macilentus* Wiedemann, 1820], by monotypy. Junior homonym of *Systropus* Wiedemann, 1820. Synonymized with *Systropus* Wiedemann, 1820, implicitly by Hagen (1862:356) and explicitly by Evenhuis (1989b:361).

Remarks. See remarks under *Apatomyza* Hennig.

Systropus Wiedemann, 1820:18. Type species: *Systropus macilentus* Wiedemann, 1820, by original designation (on plate). **Systropodinae.**

Remarks. According to Bowden (1980:386–89), there are currently 4 subgenera recognized in this genus: subg. **Systropus** s. str., subg. **Diaerops** Enderlein, 1926, subg. **Dimelopelma** Enderlein, 1926, and subg. **Teinopelmus** Bowden, 1975.

Tamerlania Paramonov, 1931:205. Type species: *Tamerlania grisea* Paramonov, 1931, by monotypy. Junior homonym of *Tamerlania* Skrjabin, 1924. Objective synonym of *Timiomyia* Evenhuis, 1978.

Tauropsis Hull, 1973:365, 368 (as subgenus of *Thyridanthrax* Osten Sacken). Type species: *Anthrax irrorellus* Klug, 1832, by original designation. Raised to generic status and synonymized with *Exhyalanthrax* Becker, 1916, herein, **new synonymy.**

Remarks. Though briefly mentioned in text, there is enough information to validate the name in Hull (1973:365). It is omitted from Hull's index, which is probably why it has not appeared in any works subsequent to Hull (1973). Its type species is treated by Zaitzev (1989:137) as a member of the genus *Exhyalanthrax,* thus the new synonymy.

Teinopelmus Bowden, 1967:137 (as subgenus of *Systropus* Wiedemann). Type species: *Systropus rugosus* Bezzi, 1924, by original designation. **Systropodinae.**

Thevenemyia Bigot, 1875a:196 (22 September) [1875b:clxxiv (22 December)]. Type species: *Thevenemyia californica* Bigot, 1875, by monotypy. Unavailable name. Incorrect original spelling. Objective synonym of *Thevenetimyia* Bigot, 1875.

Thevenetimyia Bigot, 1875a:196 (emendation of *Thevenemyia* Bigot by Bigot, 1892:325). Type species: *Thevenetimyia californica* Bigot, 1875 (as "*Thevenemyia*"), automatic. **Ecliminae.**

Remarks. Previous workers (e.g., Hall 1969; Hull 1973; Evenhuis 1989b; Zaitzev 1989) treat *Thevenetimyia* as an unjustified emendation of *Thevenemyia*. It is apparent from the original description that Bigot (1875a:196) proposed the new genus-group name for M. Thevenet. The I.C.Z.N. Code [Art. 32c(ii)] states that if there is clear evidence of an inadvertent error in the original publication, the name proposed is to be considered an incorrect original spelling and corrected as per Art. 32c. Bigot's (1892:325) modification of the spelling of *Thevenemyia* to *Thevenetimyia* is therefore a justified emendation. According to Art. 33b, the emended name takes the author and date of the original spelling.

According to Hall (1969) there are 2 subgenera currently recognized in this genus: subg. **Thevenetimyia** s. str. and subg. **Arthroneura** Hull, 1966.

Thlipsigaster Marschall, 1873:348 (unjustified emendation of *Thlipsogaster* Rondani). Type species: *Thlypsomyza castanea* Macquart, 1840, automatic. Synonymized with *Amictus* Wiedemann, 1817, by Hull (1973:284).

Thlipsogaster Rondani, 1863:72 [1864:72]. Type species: *Thlypsomyza castanea* Macquart, 1840, by subsequent designation of Bezzi (1924:4). Synonymized with *Amictus* Wiedemann, 1817, by Bigot (1892:324).

Remarks. Rondani used 2 different spellings of this genus-group name in the original publication (*Thlypsogaster* on p. 72; *Thlipsogaster* on p. 98). Bezzi (1924:4) was the 1st to revise the 2 spellings and chose *Thlipsogaster* as the correct original spelling. The 2 entries (*Thlipsogaster* and *Thlypsogaster*) in Scudder (1882a) do not satisfy the criteria of an emendation or action of 1st included revisor. Kertész (1909:87) and Schulze et al. (1938:3463) erroneously attributed authorship of *Thlipsogaster* to Coquillett (1894:108).

Thlipsomyza Henning, 1832:327. Type species: *Bombylius compressus* Fabricius, 1794, by monotypy. Junior homonym of *Thlipsomyza* Wiedemann, 1820. Synonymized with *Amictus* Wiedemann, 1817, by Hagen (1862:356).

Remarks. See remarks under *Apatomyza* Henning.

Thlipsomyza Wiedemann, 1820:12. Type species: *Bombylius compressus* Fabricius, 1805, by original designation (on plate). Synonymized with *Amictus* Wiedemann, 1817, by Schiner (1860b:67).

Remarks. Scudder (1882a:334) listed *Thlypsomyza* and *Tlipsomyza* (p. 336), but these are merely listings of subsequent misspellings of *Thlipsomyza* and are not considered here emendations.

Thlypsogaster: Rondani, 1863:72 [1864:72]. Incorrect original spelling. See *Thlipsogaster* Rondani.

Thlypsomyza: Verrall *in* Scudder, 1882a:334. See *Thlipsomyza* Wiedemann.

Thyridanthrax Osten Sacken, 1886b:113 (as subgenus of *Anthrax*), 123 (as "group" of *Anthrax*). Type species: *Anthrax selene* Osten Sacken, 1886, by subsequent designation of Coquillett (1910:615). Raised to generic status by Bezzi (1912:625). **Anthracinae: Villini.**

Tillyardomyia Tonnoir, 1927:101. Type species: *Tillyardomyia gracilis* Tonnoir, 1927, by monotypy. **Ecliminae.**

Timiomyia Evenhuis, 1978b:247 (new replacement name for *Tamerlania* Paramonov). Type species: *Tamerlania grisea* Paramonov, 1931, automatic. **Bombyliinae: Crocidiini.**

Remarks. The classification here follows Engel (1933b:178) in placing it in association with *Crocidium* Loew, 1860, and *Semiramis* Becker, 1913.

Tithonomyia† Evenhuis, 1984:160. Type species: *Protophthiria atra* Melander, 1949, by original designation. Fossil shale genus. ***Incertae sedis* (within Homoeophthalmae).**

Tlipsomyza: Verrall *in* Scudder, 1882a:336. See *Thlipsomyza* Wiedemann.

Tmemophlebia Evenhuis, 1986:19 (new replacement name for *Cyclorhynchus* Macquart). Type species: *Cyclorhynchus testaceus* Macquart, 1840, automatic. **Phthiriinae: Poecilognathini.**

Tomomyza Henning, 1832:324. Type species: *Tomomyza anthracoides* Henning, 1832 [preoccupied, = *Tomomyza anthracoides* Wiedemann, 1820], by monotypy.

Junior homonym and synonym of *Tomomyza* Wiedemann, 1820. Synonymized with *Tomomyza* by Hagen (1862:356).

Remarks. See remarks under *Apatomyza* Henning.

Tomomyza Wiedemann, 1820:9. Type species: *Tomomyza anthracoides* Wiedemann, 1820, by original designation (on plate). **Tomomyzinae.**

Tornotes Gistl, 1848:x (unjustified new replacement name for *Ploas* Latreille). Type species: *Ploas hirticornis* Latreille, 1805 [= *Bombylius virescens* Fabricius, 1787], automatic. Synonymized with *Conophorus* Meigen, 1803, herein, **new synonymy.**

Tovlinius Zaitzev, 1979:116. Type species: *Bombylius albissimus* Zaitzev, 1964, by original designation. **Bombyliinae: Dischistini.**

Toxomyia Hull, 1973:232 (as subgenus of *Toxophora* Meigen). Type species: *Toxophora maxima* Coquillett, 1886, by original designation. Junior homonym of *Toxomyia* Felt, 1911. Raised to generic status and synonymized with *Toxophora* Meigen, 1803, by Painter, Painter and Hall (1978:20).

Toxophora Meigen, 1803:270. Type species: *Toxophora maculata* Meigen, 1804 [preoccupied, = ?*Asilus fasciculatus* Villers, 1789], by subsequent monotypy in Meigen (1804:273). **Toxophorinae.**

Remarks. Scudder (1882a:338) listed *Toxopora,* but this is merely a listing of a subsequent spelling of *Toxophora* and not an emendation.

Toxophorinarum: Verrall, 1909a:479. Type species: *Cyrtophorus pictipennis* Bigot, 1857, by present designation of the 1st species included herein. Unavailable name. Proposed in synonymy with *Cyrtophorus* Bigot, 1857; not made available before 1961. Synonymized here with *Cyrtomyia* Bigot, 1892.

Toxopora: Verrall *in* Scudder, 1882a:338. See *Toxophora* Meigen.

Trinaria Mulsant, 1852:20. Type species: *Anthrax interrupta* Mulsant, 1852 [= *Anthrax rutila* Wiedemann, 1818], by subsequent designation of Coquillett (1910:617). Synonymized with *Exoprosopa* Macquart, 1840, by Bigot (1858:573).

Triodina Hull, 1973:322 (as subgenus of *Aphoebantus* Loew; new replacement name for *Triodites* Osten Sacken). Type species: *Triodites mus* Osten Sacken, 1877, automatic. Raised to generic status and synonymized with *Aphoebantus* Loew, 1872, by Painter, Painter and Hall (1978:25).

Triodites Osten Sacken, 1877:245. Type species: *Triodites mus* Osten Sacken, 1877, by monotypy. Junior homonym of *Triodites* Gistl, 1848. Synonymized with *Aphoebantus* Loew, 1872, by Coquillett (1886a:81).

Remarks. The listing of *Triodytus* Riley (1880:279) by Schulze et al. (1940:3554) is puzzling because a check of Riley's (1880) article shows that he correctly spells the name as *Triodites* and attributes authorship to Osten Sacken. I could not find the spelling "Triodytus" in any of Riley's publications mentioning Bombyliidae and assume Schulze et al. (1940) were merely following Kertész (1909:74) in listing *Triodytus* as from Riley (1880).

Triodytus: Schulze et al., 1940:3554. See *Triodites* Osten Sacken.

Triplasius Loew, 1855:7 (as subgenus of *Bombylius* Linnaeus). Type species: *Bombylius bivittatus* Loew, 1855, by monotypy. Raised to generic status by Bigot (1858:578). Synonymized with *Bombylius* Linnaeus, 1758, by Edwards (1937:33).

Triploechus Edwards, 1937:31. Type species: *Bombylius heteronevrus* Macquart, 1850, by original designation. **Bombyliinae: Dischistini.**

Tritoneura Schiner, 1867:312 [1868:133]. Type species: *Comptosia lugubris* Philippi, 1865 [preoccupied, = *Comptosia lugubris* Rondani, 1863], by original designation. Synonymized with the subgenus *Macrocondyla (Lyophlaeba)* Rondani, 1863, questionably by Mik (1882:45) and explicitly by Edwards (1934:33).

Truquia Rondani, 1863:72 [1864:72]. Type species: *Truquia insularis* Rondani, 1863, by original designation. Synonymized with *Amictus* Wiedemann, 1817, by Verrall (1909b:17).

Tumulus Hull, 1973:403 (as subgenus of *Thyridanthrax* Osten Sacken). Type species: *Anthrax misellus* Loew, 1869, by original designation. Junior homonym of *Tumulus* Monterosato, 1888. See *Veribubo* Evenhuis.

Turkmenialla: Paramonov, 1939a:43. Incorrect original spelling. See *Turkmeniella* Paramonov.

Turkmeniela: Paramonov, 1939a:46–47. Incorrect original spelling. See *Turkmeniella* Paramonov.

Turkmeniella Paramonov, 1939a:44. Type species: *Turkmeniella magnifica* Paramonov, 1939, by original designation. **Anthracinae: Anthracini.**

Remarks. Paramonov's (1939a) original description contained 3 different spellings of this genus (*Turkmeniella* on p. 43, 44; *Turkmenialla* on p. 43; *Turkmeniela* on p. 46–47). I consider *Turkmeniela* and *Turkmenialla* to be incorrect original spellings of *Turkmeniella.*

Ulosoma: Hull, 1973:62. Incorrect original spelling. See *Ulosometa* Hull.

Ulosometa Hull, 1973:359. Type species: *Ulosometa falcata* Hull, 1973, by original designation. **Lomatiinae: Comptosiini.**

Remarks. Hull used 2 different spellings for this genus in the original publication (*Ulosoma* on p. 62; *Ulosometa* on p. 359). I consider *Ulosoma* to be an incorrect original spelling of *Ulosometa.*

Unduloptomyia Evenhuis, 1978b:247 (new replacement name for *Comastes* Osten Sacken). Type species: *Comastes robustus* Osten Sacken, 1877, automatic. Synonymized with *Heterostylum* Macquart, 1848, by Hall and Evenhuis (1981:162).

Usia Latreille, 1802:430 (unjustified new replacement name for *Voluccella* Fabricius). Type species: *Volucella florea* Fabricius, 1794, by subsequent designation of Latreille (1810:444). **Usiinae: Usiini.**

Remarks. Previous listings of this genus in regional catalogs give incorrect citations as to its type species fixation. Bowden (1975:170) states the type species as *Volucella florea* by original designation and Bowden (1980:389) and Zaitzev (1989:63) list the same species as by monotypy. All apparently missed the last paragraph in Latreille (1802:430) in which he explicitly states that this name is intended to replace *Voluccella* of Fabricius. Fabricius's (1794) description of *Voluccella* included three species, thus a subsequent type designation is needed. Latreille (1810:440) appears to be the first valid type species designation for *Usia.*

Latreille (1802) proposed *Usia* as a new replacement name for *Voluccella* Fabricius, erroneously thinking it was preoccupied by *Volucella* Geoffroy, 1762. The I.C.Z.N. (1957:83–120; Opinion 441) determined, incorrectly, that *Voluccella* Fabricius was an incorrect subsequent spelling of *Volucella* Geoffroy and placed it on the Official List of Rejected and Invalid Names in Zoology. Geoffroy's *Volucella* pertained to a syrphid genus, whereas Fabricius's *Voluccella* was spelled differently, did not mention Geoffroy, and contained solely bombyliid species. Since it is not preoccupied, *Voluccella* Fabricius, 1794, takes priority over *Usia* Latreille, 1802.

Because *Usia* is more commonly used than *Voluccella*, an application is in preparation to suppress *Voluccella* in favor of *Usia*.

Velocia Coquillett, 1886b:158. Type species: *Anthrax cerberus* Fabricius, 1794, by original designation. Junior homonym of *Velocia* Robineau-Desvoidy, 1863. Synonymized with *Ligyra* Newman, 1841, by Paramonov (1939b:39).

Venallites: Cockerell, 1913:230, 231. Incorrect original spelling. See *Verrallites* Cockerell.

Veribubo Evenhuis, 1978b:247 (new replacement name for *Tumulus* Hull). Type species: *Anthrax misellus* Loew, 1869, automatic. **Anthracinae: Villini.**

Verrallites† Cockerell, 1913:231. Type species: *Verrallites cladurus* Cockerell, 1913, by monotypy. Fossil shale genus. **Anthracinae: Villini.**

Remarks. Cockerell used 2 different spellings for this genus in his original description (*Verrallites* on p. 231; *Venallites* on p. 230, 231). Since the name is dedicated to G. H. Verrall, I consider *Venallites* to be an incorrect original spelling of *Verrallites*.

Villa Lioy, 1864:732. Type species: *Anthrax concinna* Meigen, 1820 [= *Anthrax abbadon* Fabricius, 1794], by subsequent designation of Coquillett (1910:619). **Anthracinae: Villini.**

Remarks. Schulze et al. (1940:3640) listed an earlier use of *Villa* by Deshayes (1832:1134). A check of the latter work shows that it is invalid according to Opinion 21 (I.C.Z.N. 1910b) because it refers to a pre-Linnaean publication of the name by Klein, 1753.

Villoestrus Paramonov, 1931:93. Type species: *Villoestrus uvarovi* Paramonov, 1931, by monotypy. **Anthracinae: Villoestrini.**

Voluccella Fabricius, 1794:412. Type species: *Voluccella florea* Fabricius, 1794, automatic. **Usiinae: Usiini.**

Remarks. There are 2 different spellings for this genus in the original publication (*Voluccella* on p. 412; and *Volvicella* on the last unnumbered page of the generic index). I consider *Volvicella* to be an incorrect original spelling of *Voluccella*. Because *Usia* Latreille, 1802, was proposed as a new replacement name for *Voluccella* Fabricius, 1794 (even if unnecessarily), and its type species *(Voluccella florea)* was designated by Latreille (1810), the type species for *Voluccella* is the same. Contrary to listings in some nomenclators and catalogs (e.g., Scudder 1882a:351) and previous action by I.C.Z.N. (1957), *Voluccella* Fabricius, 1794, is not preoccupied by nor is it a misspelling of *Volucella* Geoffroy, 1762. See remarks under *Usia* for details on conserving *Usia* Latreille, 1802, over *Voluccella* Fabricius, 1794.

Volvicella: Fabricius, 1794: [5—last unnumbered page of index]. Incorrect original spelling. See *Voluccella* Fabricius.

Walkeromyia Paramonov, 1934a:22. Type species: *Anthrax lurida* Walker, 1857, by original designation. **Anthracinae: Anthracini.**

Xenoprosopa Hesse, 1956b:942. Type species: *Xenoprosopa paradoxa* Hesse, 1956, by original designation. **Xenoprosopinae.**

Xenox Evenhuis, 1985a:166. Type species: *Nemotelus tigrinus* De Geer, 1776, by original designation. **Anthracinae: Anthracini.**

Xeramoeba Hesse, 1956a:356. Type species: *Xeramoeba apricaria* Hesse, 1956, by original designation. **Anthracinae: Petrorossiini.**

Xethomyza: Hesse, 1938:986, 989. Type species: *Eurycarenus pachyceratus* Bigot, 1892, by monotypy. Unavailable name. Originally proposed in synonymy with *Oniromyia* Bezzi, 1921; not made available before 1961.

Xystropus Agassiz, 1847:359, 390 (unjustified emendation of *Systropus* Wiedemann). Type species: *Systropus macilentus* Wiedemann, 1820, automatic. Junior homonym of *Xystropus* Solier, 1835. Objective synonym of *Systropus* Wiedemann, 1820.

Ylasoia Speiser, 1920:213. Type species: *Anthrax pegasus* Wiedemann, 1828, by original designation. **Lomatiinae: Comptosiini.**

Zaclava Hull, 1973:250. Type species: *Systropus clavifemoratus* Hardy, 1921, by original designation. **Systropodinae.**

Zephyrectes Evenhuis, 1978a:33 (as subgenus of *Bombylius* Linnaeus). Type species: *Bombylius montanus* Johnson and Johnson, 1975, by original designation. **Bombyliinae: Bombyliini.**

Zerinthia Rafinesque, 1815:130 (unnecessary new replacement name for *Usia* Latreille). Type species: *Voluccella florea* Fabricius, 1794, automatic. Synonym of *Usia* Latreille, 1802, **new synonymy.**

Zinnomyia Hesse, 1955:387. Type species: *Zinnomyia karooensis* Hesse, 1955, by original designation. **Bombyliinae: Dischistini.**

Zygodipla Bezzi, 1924:244 (as subgenus of *Exoprosopa* Macquart) [1926:268 (as genus)]. Type species: *Anthrax algira* Fabricius, 1794, by original designation. Raised to generic status in Bezzi (1926:268). Synonymized with *Heteralonia* Rondani, 1863, by Zaitzev (1989:150).

Remarks. Though previous works (e.g., Hull 1973:413; Zaitzev 1989:150) dated this genus from its description in 1926, it was originally proposed in synonymy with *Exoptata* Coquillett, 1887, in Bezzi (1924:244) with *Anthrax algira* designated as the type species on p. 245. It was subsequently made available in Bezzi (1926:268) but dates from the 1924 work.

Zyxmyia Bowden, 1960:213. Type species: *Zyxmyia megachile* Bowden, 1960, by original designation. **Cythereinae: Mariobezziini.**

Literature Cited

Agassiz, L. 1846. Nomina systematica generum Dipterorum, tam viventium quam fossilium, secundum ordinem alphabeticum disposita, adjectis auctoribus, libris in quibus reperiuntur, anno editionis, etymologia et familiis ad quas pertinent. In: L. Agassiz, Nomenclator zoologicus continens nomina systematica generum animalium tam viventium quam fossilium, secundum ordinem alphabeticum disposita, adjectis auctoribus, libris, in quibis reperiuntur, anno editionis, etymologia et familias, ad quas pertinent, in singulis classibus, [Pt. 4], [vi] + 42 p. Fasc. IX/X: Titulum et praefationem operis, Mollusca, Lepidoptera, Strepsiptera, Diptera, Myriapoda, Thysanura, Thysanoptera, Suctoria, Epizoa et Arachnidas. Jent & Gassman, Soloduri [= Solothurn, Switzerland]. (before 25 November)
[Recorded in the 25 November issue of Bibliogr. Fr.]

——. 1847. Nomenclatoris zoologici index universalis, continens nomina systematica classium, ordinum, familiarum et generum animalium omnium, tam viventium quam fossilium, secundum ordinem alphabeticum unicum disposita, adjectis homonymiis plantarum, nec non variis adnotationibus et emendationibus. Jent & Gassman, Soloduri [= Solothurn, Switzerland]. viii + 393 p. (31 December +)

Aldrich, J. M. 1905. A catalogue of North American Diptera (or two-winged flies). Smithson. Misc. Coll. 46(2):1–680. (before 25 May)
[Recorded in the 25 May issue of Bibliogr. Zool.]

Arnaud, P. H., Jr. 1966. Date of publication of Genera Insectorum, Fascicle 185, Family Empididae by Axel Leonard Melander. Pan-Pac. Entomol. 42:337–38. (30 November)

Audouin, J. V., E. Blanchard, L. Doyère & H. Milne Edwards. [1836–1849]. Les insectes. Avec un atlas. Orthoptères, Hémiptères, Névroptères, Hyménoptères, Lepidoptères, Rhipiptères et Diptères. In: Cuvier, G.L.C.F.D., Le règne animal distribué d'après son organisation pour servir de base à l'histoire naturelle des animaux, et d'introduction à l'anatomie comparée, par Georges Cuvier. Edition accompagnée de planches gravées, représentant les types de tous les genres, les caractères distinctifs des divers groupes et les modifications de structure sur lesquelles repose cette classification; par une réunion de disciples de Cuvier. Vol. II. Atlas. Plate 168. Paris. (October)
[The Diptera, on plates 160–82, are by Blanchard, to whom type designations given there are attributed. Plate 168 was issued in October. For dating of the other plates in this series, see Cowan (1976).]

Austen, E. E. 1937. The Bombyliidae of Palestine. British Museum (Natural History), London. ix + 188 p. (24 July)

Becker, T. 1903. Die paläarktischen Formen der Gattung *Mulio* Latreille (Dipt.) [part]. Z. Syst. Hymen. Dipt. 3:17–23. (1 January)

——. 1906. Die Ergebnisse meiner dipterologischen Frühjahrsreise nach Algier und Tunis. 1906 [part]. Z. Syst. Hymen. Dipt. 6:145–58. (1 March)

——. 1907a. Die Ergebnisse meiner dipterologischen Frühjahrsreise nach Algier und Tunis. 1906 [part]. Z. Syst. Hymen. Dipt. 7:97–128. (1 February)

——. 1907b. *Legnotus* Lw. (Dipt.). Z. Syst. Hymen. Dipt. 7:452–54. (1 November)

——. 1907c. Zur Kenntniss der Dipteren von Central-Asien. 1. Cyclorrhapha Schizophora Holometopa und Orthorrhapha Brachycera. Ezheg. Zool. Muz. 12:253–317. (October)

——. 1913a. Genera Bombyliidarum. Ezheg. Zool. Muz. 17:421–502. (March)

——. 1913b. In: Becker, T. & P. Stein, Persische Dipteren von den Expeditionen des Herrn N. Zarudny 1898 und 1901. Ezheg. Zool. Muz. 12:503–654. (March)

[Evenhuis et al. (1989:824) state that Stein is mentioned on the title page as only having provided assistance and was erroneously listed in some bibliographies as a co-author. I recently obtained a separate of this paper in which Stein is listed on the wrapper as a co-author. I interpret the wrapper as giving evidence that Stein should be considered co-author of the entire work. Becker is the sole author of new taxa of Bombyliidae and other Diptera described in this paper. Stein is mentioned in a footnote (p. 618) as responsible for the descriptions of new taxa of Anthomyiidae and Tachinidae and determinations of species in those families.]

——. 1914. Diptères nouveaux récoltés par MM. Ch. Alluaud et R. Jeannel en Afrique orientale. Ann. Soc. Entomol. Fr. 83:121–30. (22 April)

——. 1915. *Edmundiella,* novum genus Lomatiinarum. Wien. Entomol. Ztg. 34:347–48. (22 October)

——. 1916. Beiträge zur Kenntniss einiger Gattungen der Bombyliiden. Ann. Mus. Natl. Hist. Nat. Hung. 14:17–67. (25 June)

——. 1920. *Conophorina,* novum genus Bombyliidarum (Dipt.). Entomol. Mitt. 9:181–84. (30 October)

Berthold, A. A. 1827. Natürliche Familien des Theirreichs. Aus dem Französischen. Mit Anmerken und Zusätzen. Landes-Industrie, Weimar. 606 p. (31 December+)

[This is a German translation of Latreille (1825) and contains Latinizations of some of Latreille's vernacular names.]

Bezzi, M. 1902. Neue Namen für einige Dipteren-Gattungen. Z. Syst. Hymen. Dipt. 2:190–92. (1 May)

——. 1903. Band II. Orthorrhapha Brachycera. In: T. Becker, M. Bezzi, J. Bischof, K. Kertész & P. Stein, eds., Katalog der paläarktischen Dipteren. Budapest. 396 p. (before 1 April)

[Recorded in the 1 April listing of Bibliogr. Zool. for 1903.]

——. 1908. Eine neue *Aphoebantus* Art aus den palaearktischen Faunengebiete (Dipt.). Z. Syst. Hymen. Dipt. 8:26–36. (1 January)

——. 1912. Report on a collection of Bombyliidae (Diptera) from central Africa, with description of new species. Trans. Entomol. Soc. Lond. 1911:605–56. (10 February)

——. 1921a. On the bombyliid fauna of South Africa (Diptera) as represented in the South African Museum. Ann. S. Afr. Mus. 18:1–180. (21 February)

——. 1921b. Additions to the bombyliid fauna of South Africa (Diptera), as represented in the South African Museum. Ann. S. Afr. Mus. 18:469–78. (6 December)

——. 1922a. Materiali per lo studio della fauna Tunisina raccolti da G. e L. Doria. Mus. Civ. Stor. Nat. Giacomo Doria 50:97–139. (22 September)

——. 1922b. Contributo allo studio della fauna Libica. Ditteri di Cirenaica. Raccolti dal Rev. Miss. Don Vito Zanon. Mem. Soc. Entomol. Ital. 1:140–57. (30 December)

——. 1922c. Enumeratio bombyliidarum (Dipt.) quas ex Africa meridionali Dr. H. Brauns misit. Brotéria 20:64–86. (1 August)

——. 1923. Insectes Diptères. Bombyliidae et Syrphidae. Result. Sci. Voy. Ch. Alluaud et R. Jeannel en Afrique orientale (1911–1912) 6:315–51. (15 April)

——. 1924. The Bombyliidae of the Ethiopian Region. British Museum (Natural History), London. 390 p. (29 April*)

——. 1925. Quelques notes sur les bombyliides (Dipt.) d'Egypte, avec decription d'espèces nouvelles. Bull. Soc. R. Entomol. Égypte 8:159–242. (31 December +)

——. 1926. Notes additionelles sur les Bombyliidae (Dipt.) d'Égypte. Bull. Soc. R. Entomol. Égypte 9:244–73. (31 December +)

Bigot, J.M.F. 1857. Diptères nouveaux provenant du Chili. Ann. Soc. Entomol. Fr. (3) 5:277–308. (9 September)

——. 1858. Essai d'une classification générale et synoptique de l'ordre des insectes diptères. VIe mémoire. Ann. Soc. Entomol. Fr. (3) 6:569–87. (24 November)

——. 1859. Essai d'une classification générale et synoptique de l'ordre des insectes diptères. VIIe mémoire. Tribus des Rhaphidi et Dolichopodi (mihi). Ann. Soc. Entomol. Fr. (3) 7:201–31. (24 August)

——. 1875a. [Description d'un nouvelle espèce de diptère]. Bull. Bimens. Soc. Entomol. Fr. 60:196–97. (30 September)

——. 1875b. [Description d'un nouvelle espèce de diptère]. Bull. Soc. Entomol. Fr. (5) 5:clxxiv–vi. (22 December)

——. 1876a. [Diagnose d'un nouveaux genre de diptères]. Bull. Bimens. Soc. Entomol. Fr. 73:71. (20 April)

——. 1876b. [Diagnose d'un nouveaux genre de diptères]. Bull. Soc. Entomol. Fr. (5) 6:lxvi. (11 October)

——. 1886a. [Diagnoses nouvelles d'un genre et d'une espèce de l'ordre de diptères]. Bull. Bimens. Soc. Entomol. Fr. 1886 (12):ciii–civ. (3 July)

——. 1886b. [Diagnoses d'un genre et d'une espèce de diptères]. Bull. Bimens. Soc. Entomol. Fr. 1886 (13):cx–cxi. (7 August)

——. 1886c. [Diagnoses nouvelles d'un genre et d'une espèce de l'ordre de diptères]. Bull. Soc. Entomol. Fr. (6) 6:ciii–civ. (15 October)

——. 1886d. [Diagnose d'un nouveaux genre et d'une espèce de diptères]. Bull. Soc. Entomol. Fr. (6) 6:cx. (15 October)

——. 1887a. [Diagnoses d'un genre nouveau et d'un espèce nouvelle de l'ordre de diptères]. Bull. Bimens. Soc. Entomol. Fr. 1887(3):xxxi. (19 February)

——. 1887b. [Remarques diptèrologiques]. Bull. Bimens. Soc. Entomol. Fr. 1887(6):lx–lxi. (3 April)

——. 1887c. [Diagnoses d'un genre nouveau et d'un espèce nouvelle de l'ordre de diptères]. Bull. Soc. Entomol. Fr. (6) 7:xxxi. (25 May)

——. 1887d. [Remarques diptèrologiques]. Bull. Soc. Entomol. Fr. (6) 7:lx–lxi. (25 May)

——. 1888. Description d'un nouveaux genre de diptère]. Bull. Bimens. Soc. Entomol. Fr. 1888 (18):cxl. (6 October)

——. 1889. [Description d'un nouveaux genre de diptère]. Bull. Soc. Entomol. Fr. (6) 8:cxl. (8 January)

——. 1892. Diptères nouveaux ou peu connus. 37ᵉ partie. XLVI. Bombylidi (mihi). Ann. Soc. Entomol. Fr. 61:321–76. (28 December)

Bischof, J. 1903. Neue Dipteren aus Afrika. Wien. Entomol. Ztg. 22:41–42. (5 March)

Blackwelder, R. E. 1967. Taxonomy. A text and reference book. John Wiley & Sons, New York. xiv + 698 p. (31 December+)

Bowden, J. 1960. Studies in African Bombyliidae. IV. A new genus of the Corsomyza-group from east Africa. J. Entomol. Soc. South. Afr. 23:211–17. (15 May)

——. 1962. Studies in African Bombyliidae. V. A new genus of Exoprosopinae from east Africa. J. Entomol. Soc. South. Afr. 25:116–20. (30 April)

——. 1964. The Bombyliidae of Ghana. Mem. Entomol. Soc. South. Afr. 8:1–159. (May)

——. 1967. Studies in African Bombyliidae. VI. A provisional classification of the Ethiopian Systropinae with descriptions of new and little known species. J. Entomol. Soc. South. Afr. 30:126–73. (30 September)

——. 1971. Notes on the genus *Ligyra* Newman (Diptera:Bombyliidae) with descriptions of three new species from the New Guinea Subregion. J. Aust. Entomol. Soc. 10:5–12. (31 March)

——. 1973. Studies in African Bombyliidae. VII. On *Dischistus* Loew and related genera, and *Bombylisoma* Rondani, with some zoogeographical considerations. J. Entomol. Soc. South. Afr. 36:139–58. (March)

——. 1974. Studies in African Bombyliidae. VIII. On the Geroninae. J. Entomol. Soc. South. Afr. 37:87–108. (March)

——. 1975a. Studies in African Bombyliidae. IX. On *Hyperusia* Bezzi and the tribe Corsomyzini. J. Entomol. Soc. South. Afr. 38:99–107. (March)

——. 1975b. Studies in African Bombyliidae. X. Taxonomic problems relevant to a catalogue of Ethiopian Bombyliidae with descriptions of new genera and species. J. Entomol. Soc. South. Afr. 38:305–20. (September)

——. 1975c. Family Bombyliidae. In: M. D. Delfinado & D. E. Hardy, eds., A catalog of the Diptera of the Oriental Region, p. 165–84. Vol. II. Suborder Brachycera through Division Aschiza, Suborder Cyclorrhapha. Univ. Press of Hawaii, Honolulu. [ix] + 459 p. (5 October)
[Dated from information supplied by the publisher.]

——. 1980. Family Bombyliidae. In: Crosskey, R. W., ed., A catalogue of the Diptera of the Afrotropical Region, p. 381–430. British Museum (Natural History), London. 1,437 p. (10 July)
[Dated from a stamped copy in the BMNH Diptera section.]

——. 1985. The tribal classification of the Bombyliinae with particular reference to the Bombyliini and Dischistini, and the description of a new genus from South America (Dipt., Bombyliidae). Entomol. Mon. Mag. 121:99–107. (29 July)

Brèthes, J. [1926]. Sur quelques dipteres chiliens. Rev. Chil. Hist. Nat. 28:104–11. [In French.] (20 August*)

[Though the title page has the year 1924, and Perez (1968:135) dates it as 1924, the volume contains reference to papers in 1925 and was received by BMNH on 20 August 1926. I accept 1926 as the actual year of publication. Actual date of issue was probably a month or two earlier than the BMNH receipt date.]

Brues, C. T. & A. L. Melander. 1932. Classification of insects. A key to the known families of insects and other terrestrial arthropods. Bull. Mus. Comp. Zool. 73:1–672. (January)

Brues, C. T., A. L. Melander & F. M. Carpenter. 1954. Classification of insects. Keys to the living and extinct families of insects, and to the living families of other terrestrial arthropods. Bull. Mus. Comp Zool. 108:i–v, 1–917. (March)

Brunetti, E. 1909. Revised and annotated catalogue of Oriental Bombyliidae with descriptions of new species. Rec. Indian Mus. 2:437–92. (January)

——. 1920. Diptera Brachycera. Vol. I. In: A. E. Shipley, ed., The fauna of British India, including Ceylon and Burma. Taylor and Francis, London. ix + 401 p. (May)

Carrera, M. & M.A.V. d'Andretta. 1950. Sobre as espécies Brasileiras de *Systropus* Wiedemann, 1830 (Diptera, Bombyliidae). Pap. Avulsos Dep. Zool. 9:295–319. (15 November)

Cockerell, T.D.A. 1909a. Fossil Diptera from Florissant, Colorado. Bull. Am. Mus. Nat. Hist. 26:9–12. (19 January)

——. 1909b. Descriptions of Tertiary insects. Part 5. Some new Diptera. Am. J. Sci. (4) 27:53–58. (January)

——. 1909c. Fossil insects from Florissant, Colorado. Bull. Am. Mus. Nat. Hist. 26:67–76. (9 March)

——. 1913. Some fossil insects from Florissant, Colorado. Can. Entomol. 45:229–33. (8 July)

——. 1914a. The fossil and recent Bombyliidae compared. Bull. Am. Mus. Nat. Hist. 33:229–36. (6 April)

——. 1914b. New and little-known insects from the Miocene of Florissant, Colorado. J. Geol. 22:714–24. (November)

——. [1915]. Miocene fossil insects. Proc. Acad. Sci. Phila. 66:634–48. (23 January)

[Though commonly dated as 1914, the mailing date of the issue containing this article was actually 23 January 1915.]

——. 1916. Some American fossil insects. Proc. U.S. Natl. Mus. 51:89–106. (16 October)

——. 1917. New Tertiary insects. Proc. U.S. Natl. Mus. 52:373–84. (23 February)

Cole, F. R. 1923. A revision of the North American two-winged flies of the family Therevidae. Proc. U.S. Natl. Mus. 62(4):1–140. (10 May)

Coquillett, D. W. 1886a. Monograph of the Lomatina of North America. Can. Entomol. 18:81–87. (22 July)

——. 1886b. The North American genera of Anthracina. Can. Entomol. 18:157–59. (29 October)

——. 1887. Notes on the genus *Exoprosopa*. Can. Entomol. 19:12–14. (14 March)

——. 1891. New Bombyliidae of the group *Paracosmus*. West Am. Sci. 7:219–22. (September)

[Published simultaneously with same title in Science and Horticulture, volume 1, number 5.]

——. 1893. An anomalous empid. Entomol. News 4:208–10. (June)

——. 1894. Notes and descriptions of North American Bombyliidae. Trans. Am. Entomol. Soc. 21:89–112. (March)

——. 1910. The type-species of the North American genera of Diptera. Proc. U.S. Natl. Mus. 37:499–647. (4 August)

Costa, A. 1863. Nuovi studie sulla entomologia della Calabria Ulteriore. Atti Accad. Sci. Fis.-Mat. Napoli 1(2):1–80. (31 December +)

Coucke, E. & L. Coucke. 1894. Matériaux pour une étude des diptéres de Belgique. Bombylides. Ann. Soc. Entomol. Belg. 38:281–92. (31 December +)

Cowan, C. F. 1970. Anotationes Ropalocerologicae 1970. Kettering, U.K. 70 p. (31 December +)

Cowan, C. F. 1976. On the Disciples' Edition of Cuvier's Règne animal. J. Soc. Bibliogr. Nat. Hist. 8:32–64. (November)

Cresson, E. T. 1915. A new genus and some new species belonging to the dipterous family Bombyliidae. Entomol. News 26:200–207. (30 April)

Curran, C. H. 1934. The families and genera of North American Diptera. Ballou Press, New York. 512 p. (25 August)

Dallas, W. S. 1865. Insecta. In: A.C.L.G. Günther, ed., The record of zoological literature. 1864, p. 328–589. Volume first. Zoological Society, London. 634 p. (25 September) [Dated from an advertisement on the last page of the volume.]

——. 1866. Insecta. In: A.C.L.G. Günther, ed., The record of zoological literature. 1865, p. 381–710. Volume second. Zoological Society, London. 798 p. (before November) [Reviewed in the November issue of Entomol. Mon. Mag.]

Deshayes, G. P. 1832. Encyclopédie méthodique. Histoire naturelle des vers. Tome troisième. Agasse, Paris. P. 595–1151. (September) [Dated from Sherborn and Woodward (1899:595).]

Edwards, F. W. 1930. Bombyliidae, Nemestrinidae and Cyrtidae. In: Diptera of Patagonia and south Chile based mainly on material in the British Museum (Natural History), p. 62–197. Part V. Fasc. 2. British Museum (Natural History), London. (26 July)

——. 1934. On the genus *Comptosia* and its allies (Bombyliidae). Encycl. Entomol. B (II) Dipt. 7:81–112. (27 September) [For dating see Evenhuis (1989a:210).]

——. 1937. The Bombyliidae of Chile and western Argentina. Rev. Chil. Hist. Nat. 40:31–46. (23 June*)

Edwards, M. A. & A. T. Hopwood. 1966. Nomenclator zoologicus. Vol. VI. 1946–1955. Zoological Society of London, London. 329 p. (31 December +)

Edwards, M. A. & H. G. Vevers. 1975. Nomenclator zoologicus. Vol. VII. 1956–1965. Zoological Society of London, London. 374 p. (31 December +)

Efflatoun, H. C. 1945. A monograph of Egyptian Diptera. Part VI. Family Bombyliidae. Section I: Subfamily Bombyliidae Homeophthalmae. Bull. Soc. Fouad Ier Entomol. 29:1–483. (31 December)

Enderlein, G. 1926. Zur Kenntnis der Bombyliiden-Subfamilie Systropodinae (Dipt.). Wien. Entomol. Ztg. 43:69–92. (10 September)

——. 1934. Entomologische Ergebnisse der deutsch russischen Alai-Pamir Expedition, 1928. III. 1. Diptera. Dtsch. Entomol. Z. 1933:129–46. (24 February)

Engel, E. O. 1932a. Bombyliidae. In: E. Lindner, ed., Die Fliegen der palaearktischen Region, p. 1–48. Vol. 4, pt. 3. E. Schweizerbart, Stuttgart. (7 September)
[The entire work was issued in 12 lieferungen from 1932 to 1937 as follows: p. 1–48 (7 September 1932); p. 49–96 (2 November 1932); p. 97–144 (1 February 1933); p. 145–92 (20 December 1933); p. 193–256 (4 September 1934); p. 257–304 (6 February 1935); p. 305–52 (13 March 1935); p. 353–400 (9 May 1935); p. 401–48 (18 May 1936); p. 449–512 (8 September 1936); p. 513–60 (8 December 1936); p. 561–619 (26 July 1937). Dated from a tip-in sheet bound with the completed series in 1937.]

——. 1932b. Bombyliidae. In: Lindner, E., ed., Die Fliegen der palaearktischen Region, p. 49–96. Vol. 4, pt. 3. E. Schweizerbart, Stuttgart. (2 November)

——. 1933a. Bombyliidae. In: Lindner, E., ed., Die Fliegen der palaearktischen Region, p. 97–144. Vol. 4, pt. 3. E. Schweizerbart, Stuttgart. (1 February)

——. 1933b. Bombyliidae. In: Lindner, E., ed., Die Fliegen der palaearktischen Region, p. 145–92. Vol. 4, pt. 3. E. Schweizerbart, Stuttgart. (20 December)

——. 1935. Bombyliidae. In: Lindner, E., ed., Die Fliegen der palaearktischen Region, p. 305–52. Vol. 4, pt. 3. E. Schweizerbart, Stuttgart. (15 March)

——. 1936a. Bombyliidae. In: Lindner, E., ed., Die Fliegen der palaearktischen Region, p. 401–48. Vol. 4, pt. 3. E. Schweizerbart, Stuttgart. (18 May)

——. 1936b. A new genus and a new species of Bombyliidae (Dipt.) from southern Africa. Occas. Pap. Rhodes. Mus. 5:39–41. (30 May)

——. 1937. Bombyliidae. In: Lindner, E., ed., Die Fliegen der palaearktischen Region, p. 561–619. Vol. 4, pt. 3. E. Schweizerbart, Stuttgart. (26 July)

Evenhuis, N. L. 1978a. New species and a new subgenus of *Bombylius* (Diptera: Bombyliidae). Entomol. News 89:33–38. (13 March)

——. 1978b. Homonymy notes in the Bombyliidae (Diptera). III. Entomol. News 89:247–48. (30 November)

——. 1979a. Studies in Pacific Bombyliidae (Diptera). II. Revision of the genus *Geron* of Australia and the Pacific. Pac. Insects 21:13–55. (30 August)

——. 1979b. A new genus of Anthracinae (Diptera: Bombyliidae) from India. Pac. Insects 20:362–64. (25 July)

——. 1980. Studies in Pacific Bombyliidae (Diptera). V. Notes on the *Comptosia* group of the Australian Region, with key to genera and descriptions of a new genus and three new species. Pac. Insects 21:328–34. (21 March)

——. 1981. Studies in Pacific Bombyliidae (Diptera). VI. Description of a new anthracinae genus from the western Pacific, with notes on some of Matsumura's *Anthrax* types. Pac. Insects 23:189–200. (23 June)

——. 1982. New east Asian *Systropus* (Diptera: Systropodidae). Pac. Insects 24:31–38. (30 April)

——. 1983a. Studies in Pacific Bombyliidae (Diptera). IX. Systematic remarks on Australian Bombyliinae, with descriptions of new genera. Int. J. Entomol. 25:206–14. (27 October)

——. 1983b. An indexed bibliography of Bombyliidae (Insecta, Diptera). J. Cramer, Braunschweig [= Brunswick]. 493 p. (15 December)
[Dated from correspondence with the publisher.]

——. 1983c. Nomenclatorial notes on the genera *Robertsella* and *Meganthrax* (Diptera: Bombyliidae). Int. J. Entomol. 25:324. (29 December)

——. 1984. A reassessment of the taxonomic position of the fossil species *Protophthiria palpalis* and *P. atra* (Diptera: Bombyliidae). Int. J. Entomol. 26:157–61. (29 March)

——. 1985a. The status of the genera of the tribe Anthracini (Diptera: Bombyliidae). Int. J. Entomol. 27:162–69. (19 April)

——. 1985b. Replacement name for *Syrphoides* Evenhuis (Diptera: Bombyliidae). Int. J. Entomol. 27:289. (10 September)

——. 1986. The genera of the Phthiriinae of Australia and the New World (Diptera: Bombyliidae). Privately published, Honolulu. 57 p. (15 April)

——. 1989a. Dating of Encyclopédie Entomologique Série B. II. Diptera. Arch. Nat. Hist. 16:209–11. (June)

——. 1989b. Family Bombyliidae. In: N. L. Evenhuis, ed., Catalog of the Diptera of the Australasian and Oceanian Regions, p. 359–74. E. J. Brill, Leiden & Bishop Museum Press, Honolulu. 1,155 p. (23 August)

——. [1990]a. Systematics and evolution of the subfamilies Usiinae and Phthiriinae (Diptera: Bombyliidae) of the world. Entomonograph 11:1–78. (18 January)
[Title page states also published as Bishop Mus. Bull Entomol. 3. The title page has the date 1989, but the publication was not mailed until 18 January 1990.]

——. 1990b. Dating of the livraisons and volumes of d'Orbigny's Dictionnaire Universel d'Histoire Naturelle. Bishop Mus. Occas. Pap. 30:219–25. (6 June**)
[The inside cover of the publication lists the publication date as 8 June; however, copies were received at the Bishop Museum library on 6 June.]

——. 1991a. Studies in Pacific Bombyliidae (Diptera). 10. The Bombyliidae of New Caledonia. Mém. Mus. Natl. Hist. Nat., Paris (Zool.) 149:279–88. (20 February)

——. 1991b. The generic concepts of *Dischistus* Loew and *Bombylisoma* Rondani revisited, with discussion of the status of South American "Bombylius." Bishop Mus. Occas. Pap. 31:163–165.

Evenhuis, N. L. & D. J. Greathead. 1989. Case 2964. *Exoprosopa* Macquart, 1840 (Insecta, Diptera): proposed confirmation of *Anthrax pandora* Fabricius, 1805 as the type species. Bull. Zool. Nomen. 46:250. (19 December)

Evenhuis, N. L. & F. C. Thompson. 1990. Type designations of genus-group names of Diptera given in d'Orbigny's Dictionnaire Universel d'Histoire Naturelle. Bishop Mus. Occas. Pap. 30:226–58. (6 June**)
[See annotation under Evenhuis 1990b for dating information.]

Evenhuis, N. L., F. C. Thompson, A. R. Pont & B. L. Pyle. 1989. Literature cited. In: N. L. Evenhuis, ed., Catalog of the Diptera of the Australasian and Oceanian Regions, p. 809–991. E. J. Brill, Leiden & Bishop Museum Press, Honolulu. 1,155 p. (23 August)

Fabricius, J. C. 1794. Entomologia systematica emendata et aucta. Secundum classes, ordines, genera, species, adjectis synonimis, locis observationibus, descriptionibus. Tome 4. Hafniae [= Copenhagen]. [6] + 472 = [5] p. (31 December+)

——. 1805. Systema antliatorum secundum ordines, genera, species adiectus synonymis, locis, observationibus, descriptionibus. C. Reichard, Brunsvigae [= Brunswick]. xiv + p. 15–372 + [1] + 30 p. (31 December +)

Fletcher, D. S. & I.W.B. Nye. 1982. The generic names of moths of the world. Volume 4. Bombycoidea, Castinoidea, Cossoidea, Mimallonoidea, Sesoidea, Sphingoidea, Zygeanoidea. British Museum (Natural History), London. xiv + 192 p. (31 December +)

François, F. J. 1954. Contribution à l'étude des diptères de l'Urundi. VI.—Synonymie du genre *Isotamia* Bezzi (Bombyliidae) et de son génotype. Ann. Soc. Entomol. Belg. 90:180–82. (31 August)

——. 1964a. Bombyliidae (Diptera) du Musé Royal de l'Afrique centrale. I. *Palintonus,* un genre nouveau des Toxophorinae. Ann. Soc. R. Entomol. Belg. 100:323–29. (July)

——. 1964b. Récoltes de M. A. Villiers dans les dunes côtières du Sénégal (1961). Diptères Bombyliidae. Bull. Inst. Fr. Afr. Noire (A) 26:924–43. (31 December +)

——. 1967. On the generic position of « *Plesiocera* » *flavifrons* Becker and of a few related species (Diptera, Bombyliidae). Bull. Inst. R. Sci. Nat. Belg. 43(39):1–8. (31 December)

Germar, E. F. 1849. Ueber einige Insekten aus Tertiärbildungen. Dtsch. Geol. Ges. Z. 1:52–66. (31 December +)

Gistl, J. 1848. Naturgeschichte des Theirreichs, für höhere Schulen. Stuttgart. xvi + 216 p. (31 December +)

Greathead, D. J. 1967. The Bombyliidae (Diptera) of northern Ethiopia. J. Nat. Hist. 1:195–284. (28 June)

——. 1980a. Bee flies (Bombyliidae: Diptera) from Oman. J. Oman Stud., Spec. Rep. 2:233–50. (before August)
[Dated from information supplied by the author.]

——. 1980b. Insects of Saudi Arabia. Diptera: Fam. Bombyliidae. In: Wittmer, W. & W. Büttiker, eds., Fauna of Saudi Arabia, p. 291–337. Vol. 2. Pro Entomologica, Basle, Switzerland. 443 + 5 p. (15 August)

——. 1981. The *Villa* group of genera in Africa and Eurasia with a review of the genera comprising *Thyridanthrax* sensu Bezzi 1924 (Diptera: Bombyliidae). J. Nat. Hist. 15:309–26. (6 February)
[Dated from correspondence with the publisher.]

——. 1988a. Diptera: Fam. Bombyliidae of Saudi Arabia (part 2). In: Wittmer, W. & W. Büttiker, eds., Fauna of Saudi Arabia, p. 90–113. Vol. 9. Pro Entomologica, Basle, Switzerland. [4] + 477 + [3] p. (31 May)

——. 1988b. The relationship of *Tillyardomyia* Tonnoir with a redefinition of the subfamily Ecliminae (Diptera: Bombyliidae). N.Z. Entomol. 11:14–21. (June)

Greene, C. T. 1921. A new genus of Bombyliidae (Diptera). Proc. Entomol. Soc. Wash. 23:23–24. (10 February)

——. 1924. New species of *Mythicomyia* and its relationship with a new genus (Diptera). Proc. Entomol. Soc. Wash. 26:60–64. (19 March)

Griffini, A. 1896. Antracidi del Piemonte. Studio monografico. Ann. Accad. Agric. Torino 39:1–50. (before 10 August)
[Received at the 10 August meeting of the Acad. Sci. Paris.]

Guérin-Méneville, F. E. 1829. [Stygides]. In: Bory de Saint-Vincent, J.B.G.M., ed., Dictionnaire classique d'histoire naturelle, par Messieurs Audouin, Isid. Bourdon, Ad. Brogniart, De Candolle, Daudebard de Férusac, A. Desmoulins, Drapiez, Edwards, Flourens, Geoffroy de Saint-Hilaire, A. De Jussieu, Kunth, G. de Lafosse, Lamouroux, Latreille, Lucas fils, Presle-Duplessis, C. Prévost, A. Richard, Thiébaut de Barneaud, et Bory de Saint-Vincent. Ouvrage dirigé par ce dernier collaborateur, et dans lequel on a ajouté, pour le porter au niveau de la science, un grand nombre de mots qui n'avaient pu faire partie de la plupart de dictionnaires antérieurs. Tome quinzième, p. 692–93. Rey et Gravier, Paris. (May)

[See Stafleu & Cowan (1976:285–86) for dates of other volumes of this dictionnaire.]

Hagen, H. A. 1862. Bibliotheca entomologica. Die Litteratur über das ganze Gebeit der Entomologie bis zum Jahre 1862. Erster Band. A—M. W. Engelmann, Leipzig. 566 p. (before May)

[Recorded in the May issuance of Dtsch. Entomol. Z.]

Hall, J. C. 1969. A review of the subfamily Cylleniinae with a world revision of the genus *Thevenemyia* Bigot (*Eclimus* auct.) (Diptera: Bombyliidae). Univ. Calif. Publ. Entomol. 56:1–85. (25 July)

——. 1972. New North American Heterotropinae (Diptera: Bombyliidae). Pan-Pac. Entomol. 48:37–50. (5 May)

——. 1973. New Argentine *Lyophlaeba* (Diptera: Bombyliidae). Entomol. News 84:149–56. (18 September)

[All issues for 1973 were delayed by several months from their printed dates (H. Boyd, pers. comm.). This issue has an 18 September stamp in the Phila. Acad. Nat. Sci. Library.]

——. 1975a. New southwestern Bombyliidae (Diptera). Entomol. News 86:107–13. (23 October)

——. 1976a. The Bombyliidae of Chile. Univ. Calif. Publ. Entomol. 76:1–278. (March)

[Actually issued by Univ. Calif. Press in March 1976 (pers. comm.).]

——. 1976b. A revision of the North and Central American species of *Lepidanthrax* Osten Sacken (Diptera: Bombyliidae). Trans. Am. Entomol. Soc. 102:289–371. (15 December)

——. 1981. A review of the North and Central American species of *Paravilla* Painter (Diptera: Bombyliidae). Univ. Calif. Publ. Entomol. 92:i–x, 1–190, [191–99]. (January)

Hall, J. C. & N. L. Evenhuis. 1980. Family Bombyliidae. In: G.C.D.Griffiths, ed., Flies of the Nearctic Region, p. 1–96. Vol. V. Part 13. No. 1. E. Schweizerbart, Stuttgart. (June)

——. 1981. Family Bombyliidae. In: G.C.D. Griffiths, ed., Flies of the Nearctic Region, p. 97–184. Vol. V, Part 13, No. 2. E. Schweizerbart, Stuttgart. (11 November*)

——. 1986. Family Bombyliidae. In: G.C.D. Griffiths, ed., Flies of the Nearctic Region, p. 321–592. Vol. V, Part 13, No. 5. E. Schweizerbart, Stuttgart. (21 July*)

——. 1987. Family Bombyliidae. In: G.C.D. Griffiths, ed., Flies of the Nearctic Region, p. 593–656. Vol. V, Part 13, No. 6. E. Schweizerbart, Stuttgart. (31 July)

[Dated from postmark of copy sent from publisher to the authors.]

——. In press. Family Bombyliidae. In: F. C. Thompson, ed., Systematic database of Nearctic Diptera. A catalog of names of flies, mosquitoes, gnats, midges and maggots, with information on their classification, distribution and documentation. U.S. Dep. Agric., Agric. Handb.

Handlirsch, A. 1908. Die Fossilen Insekten und die Phylogenie der rezenten Formen. Ein Handbuch für Paleontologen und Zoologen. Part 7. Leipzig. P. 961–1120. (31 December +)

Hardy, G. H. 1921. Australian Bombyliidae and Cyrtidae (Dipt.). Pap. Proc. R. Soc. Tas. 1921:41–83. (20 September)

Hennig, W. 1966. Bombyliidae im Kopal und im Baltischen Bernstein (Diptera: Brachycera). Stuttg. Beitr. Naturkd. 166:1–20. (1 November)

——. 1967. Therevidae aus dem Baltischen Bernstein mit einigen Bermerkungen über Asilidae und Bombyliidae (Diptera Brachycera). Stuttg. Beitr. Naturkd. 176:1–14. (20 July)

——. 1969. Kritische Betrachtungen über die phylogenetische Bedeutung von Bernsteinfossilen: die Gattungen *Proplatypygus* (Diptera, Bombyliidae) und *Palaeopsylla* (Siphonaptera). Mem. Soc. Entomol. Ital. 48:57–67. (30 December)

——. 1973. 31. Diptera (Zweiflügler). In: J.-G. Helmcke, D. Starck & H. Wermuth, eds., Handbuch der Zoologie. Ein Naturgeschichte der Stämme des Tierreiches. IV. Band: Arthropoda—2. Hälfte: Insecta. Zweite Auflage. 2. Teil: Spezielles. De Gruyter, Berlin. 337 p. (October)

Henning, J. 1832. Nova dipterorum genera offert illustratque. Bull. Imp. Nat. Sci. Moscou 4:313–42. (28 June)
[This work is a plagiarized copy of Wiedemann, 1820, with all taxa attributed to Henning. The only features of this copy that differ from Wiedemann's is the different spelling of *Chiromyza* and the absence of the plate with type designations.]

Hermann, F. 1907. Einige neue Bombyliiden der palaearktishen Fauna (Dipt.). Z. Syst. Hymen. Dipt. 7:193–202. (1 March)

——. 1909. Diptera aus der Sinaihalbinsel. In: A. Kneucker, Zoologische Ergebnisse zweier in den Jahren 1902 und 1904 durch die Sinaihalbinsel unternommener botanischer Studienreisen, nebst zoologischen Beobachtungen aus Äegypten, Palästine und Syrien. Verh. Naturw. Ver., Karlsruhe 21:147–60. (31 December +)

Hesse, A. J. 1938. A revision of the Bombyliidae (Diptera) of Southern Africa. [I.] Ann. S. Afr. Mus. 34:1–1053. (June)

——. 1952. Contributions à l'étude des diptères de l'Urundi. II. A new genus of Bombyliidae from the Belgian Congo. Bull. Inst. Sci. Nat. Belg. 28(42):1–7. (June)

——. 1955. Diptera: Bombyliidae. S. Afr. Anim. Life 2:382–401. (31 December +)

——. 1956a. A revision of the Bombyliidae (Diptera) of Southern Africa. II [part]. Ann. S. Afr. Mus. 35:1–464. (6 September)

——. 1956b. A revision of the Bombyliidae (Diptera) of Southern Africa. II [concl.]. Ann. S. Afr. Mus. 35:465–972. (31 October)

——. 1965. Diptera (Brachycera) Bombyliidae, Cyrtosiinae. S. Afr. Anim. Life 11:482–84. (31 December +)

——. 1967. Additions to the Cyrtosiinae (Bombyliidae) of South Africa. Ann. S. Afr. Mus. 50:89–130. (December)

Higgins, L. G. 1963. Dates of publication of the Novara Reise. J. Soc. Bibliogr. Nat. Hist. 4:153–59. (November)

Hull, F. M. 1966. Notes on the genus *Diplocampta* Curran and certain other Bombyliidae. Part I. Entomol. News 77:225–27. (12 October)

——. 1970. Some new genera and species of bee flies from South America (Diptera: Bombyliidae). J. Ga. Entomol. Soc. 5:163–66. (July)

——. 1971a. Some new genera and species of bee flies (Diptera: Bombyliidae) from South America and Australia. J. Ga. Entomol. Soc. 6:1–7. (January)

——. 1971b. A new genus and new species of bee flies (Diptera: Bombyliidae). Proc. Entomol. Soc. Wash. 73:181–83. (28 June)

——. 1973. Bee flies of the world. The genera of the family Bombyliidae. Bull. U.S. Natl. Mus. 286:1–687. (12 November)

International Commission on Zoological Nomenclature. 1910a. Opinion 11. The designation of genotypes by Latreille, 1810. In: Opinions rendered by the International Commission on Zoological Nomenclature. Opinions 1 to 25. Smithson. Publ. 1938:17–18. (July)

——. 1910b. Opinion 21. Shall the genera of Klein, 1744, reprinted by Walbaum, 1792, be accepted? In: Opinions rendered by the International Commission on Zoological Nomenclature. Opinions 1 to 25. Smithson. Publ. 1938:51–52. (July)

——. 1957. Opinion 441. Validation under the Plenary Powers of the names for five genera in the Order Diptera (Class Insecta) published in 1762 by Geoffroy (E. L.) in the work entitled Histoire abrégée des Insectes qui se trouvent aux environs de Paris (Opinion supplementary to Opinion 228). Bull. Zool. Nomencl. 15:83–120. (29 January)

——. 1985. International code of zoological nomenclature. Third edition adopted by the XX General Assembly of the International Union of Biological Sciences. International Trust for Zoological Nomenclature & British Museum (Natural History), London; Univ. California Press, Berkeley, Los Angeles. 338 p. (February)

——. 1989. Opinion 1545. *Glabellula* Bezzi, 1902 (Insecta, Diptera): *Platygaster arcticus* Zetterstedt, 1838 designated as the type species. Bull. Zool. Nomen. 46:148. (23 June)

Irwin, M. E. 1976. Morphology of the terminalia and known oviposition behaviour of female Therevidae (Diptera: Asiloidea), with an account of correlated adaptations and comments on phylogenetic relationships. Ann. Natal Mus. 22:913–35. (30 November)

Irwin, M. E. & L. Lyneborg. 1981. Therevidae. [Chapter] 37. In: J. F. McAlpine, B. V. Peterson, G. E. Shewell, H. J. Teskey, J. R. Vockeroth & D. M. Wood, coordinators, Manual of Nearctic Diptera, p. 513–23. Vol. 1. Research Branch, Agriculture Canada Monograph No. 27. Ottawa. vi + 674 p. (31 March)

——. 1989. Family Therevidae. In: N. L. Evenhuis, ed., Catalog of the Diptera of the Australasian and Oceanian Regions, p. 353–58. Bishop Museum Press, Honolulu & E. J. Brill, Leiden. 1,155 p. (23 August)

Jaennicke, J. F. 1867. Neue exotische Dipteren. Abh. Senckenberg. Naturforsch. Ges. 6:311–408. (November)

Johnson, D. E. & L. M. Johnson. 1959. Taxonomic notes on North American bee flies, with descriptions of new species (Diptera: Bombyliidae). Gr. Basin Nat. 19:67–74. (31 December)

Kapoor, V. C., M. L. Agarwal & J. S. Grewel. 1979. On a collection of bombyliids (Diptera: Bombyliidae) from India. Oriental Insects 12:403–18. (January)

Karsch, F. A. 1880. Die Spaltung der Dipterngattung *Systropus* Wiedemann. Z. Ges. Naturw. 53:654–58. (July/August)

——. 1886. Dipteren von Pungo-Adongo gesammelt von Herrn Major A. von Homeyer [part]. Entomol. Nachr. 12:49–58. (February)

Kertész, K. 1909. Catalogus dipterorum hucusque descriptorum. Volumen V. Bombyliidae, Therevidae, Omphralidae. Budapest. 199 p. (before September)
[Recorded in the September issue of Nat. Novit.]

Kovalev, V. G. 1985. (Family Bombyliidae). In: N. S. Kalugina & V. G. Kovalev, eds., (Dipterous insects of the upper Jurassic), p. 186–87. Nauka, Leningrad. 196 + [3] p. (In Russian.) (2 October)

Kröber, O. 1914. Beiträge zur Kenntnis der Thereviden und Omphraliden. Mitt. Naturhist. Mus. Hamb. 31(2 Beih.):29–74. (before December)
[Recorded in the December issue of Entomol. Litt. Bl.]

Künckel d'Herculais, J. P. 1905. Les lépidoptères limacodides et leurs diptères parasites, bombylides du genre *Systropus*. Adaptation parallèle de l'hôte et du parasite aux mêmes conditions d'existence. Bull. Sci. Fr. Belg. 39:141–51. (29 July)

Latreille, P. A. 1796. Précis des caractéres génériques des insectes, disposés dans un ordre naturale. Prévôt, Paris; Brive, Bourdeaux. xiv + 201 + [7] p. (between 22 September and 31 December)
[Dated from Latreille, 1804, Tabl. Méth., p. 1, which states that this work appeared at the beginning of "l'an V." The fifth year of the French Republic began on 22 September.]

——. [1802]. Histoire naturelle, générale et particulière des Crustacés et des Insectes. Tome troisième. Familles naturelles et genres. Paris. 467 p. (6 November)
[See Fletcher & Nye (1982:xii) for dating details of this work.]

——. 1804. Tableau méthodique des insectes. In: Société de Naturalistes et Agriculteurs, Nouveau dictionnaire d'histoire naturelle, appliquée aux arts, principalement à l'agriculture et à l'économie rurale et domestique, p. 129–200. Tome XXIV. [Section 3:] tableaux méthodiques d'histoire naturelle. Déterville, Paris. 84 + 4 + 85 + 238 + 18 + 34 p. (7 March)
[Dated from Whitley (1936:189).]

——. [1805]. Histoire naturelle, générale et particulière, des crustacés et des insectes. Tome quatorzième. Ouvbrage faisant suite à l'histoire naturelle générale et particulière, composée par Leclerc de Buffon, ed rédigée par C.S. Sonnini, membre de plusieurs Sociétés savantes. Familles naturelles des genres. Dufart, Paris. 432 p. (20 February to 21 March)

——. 1810. Considérations générales sur l'ordre des animaux composant les classes des crustacés, des arachnides, et des insectes; avec un tableau méthodique de leurs genres, diposés en familles. F. Schoell, Paris. 444 p. (23 May)
[Recorded in the 23 May issue of Moniteur Universel.]

——. 1825. Familles naturelles du règne animal, exposées succinctement et dans un ordre analytique, avec l'indication de leurs genres. J.-B. Baillière, Paris. 570 p. (21 May)
[Recorded in the 21 May issue of Bibliogr. Fr.]

——. 1829. Suite et fin des insectes. In: [G.L.C.F.D.] Cuvier, Le règne animal distribué d'après son organisation, pour servir de base à l'histoire naturelle des animaux et d'intro-

duction à l'anatomie comparée. Nouvelle édition, revue et augmentée. Tome V. Déterville et Crochard, Paris. xxiv + 556 p. (11 April)
[Recorded in the 11 April issue of Bibliogr. Fr.]

Linnaeus, C. 1758. Systema naturae per regna tria naturae, secundum classes, ordines, genera, species, cum characteribus, differentiis, synonymis, locis. Tomus I. Editio decima, reformata. L. Salvii, Holmae [= Stockholm]. 824 p. (1 January)

Lioy, P. 1864. I ditteri distributi secundo un nuovo metodo di classificazione naturale [part]. Atti R. Ist. Veneto Sci., Lett. Arti (3) 9:719–71. (before 29 May)
[Recorded in the 29 May meeting of the Istituto Veneto di Scienze, Letteri et Arti.]

Loew, H. 1844. Beschreibung einiger neuen Gattungen der europäischen Dipterenfauna. Stett. Entomol. Ztg. 5:114–30, 154–73.
[Publication split: p. 114–30 (April); p. 154–73 (May).]

——. 1855. Neue Beiträge zur Kenntniss der Dipteren. Dritter Beitrag. Schulprogramm, Mezeritz. 52 p. (31 December +)

——. 1856. Neue Beiträge zur Kenntniss der Dipteren. Vierter Beitrag. Schulprogramm, Mezeritz. 57 p. (31 December +)

——. 1860a. Bidrag till kännedomen om Afrikas Diptera [part]. Öfvers. K. Vetenskapsakad. Forh. 17:81–97. (before March)
[Mentioned in the list of books published by the Academy for the year ending March 1860.]

——. 1860b. Die Dipteren-Fauna Sudafrika's. Erste Abteilung. Abh. Naturwiss. Ver. Sachsen Thüringen, Halle 2:57–402. (31 December +)
[Also published separately, 1860, with pagination xi + 330.]

——. [1869]a. Cilische Dipteren und einige mit ihren concurrirende Arten. Berl. Entomol. Z. 12:369–86. (March)

——. 1869b. Diptera Americae septentrionalis indigena. Centuria octava. Berl. Entomol. Z. 13:1–52. (May)
[Also reprinted in Loew, 1872b, p. 115–66.]

——. 1869c. Beschreibungen europäischer Dipteren. Von Johann Wilhelm Meigen. Erster Band. Systematische Beschreibung der bekannten europäischen zweiflügeligen Insecten. Achter Theil oder zweiter Supplementband. H.W. Schmidt, Halle. xvi + 310 + [1] p. (31 December +)

——. 1872a. Diptera Americae septentrionalis indigena. Centuria decima. Berl. Entomol. Z. 16:49–124. (March)
[Also reprinted in Loew, 1872b, p. 225–300.]

——. 1872b. Diptera Americae septentrionalis indigena. II. [Centuria 6–10.] A. W. Schadii, Berolini [= Berlin]. 300 p. (31 December +)
[This is a reprinted version of Loew's second 5 "centuries" with different pagination than the original journal version.]

——. 1873. Beschreibungen europäischer Dipteren. Von Johann Wilhelm Meigen. Dritter Band. Systematische Beschreibung der bekannten europäischen zweiflügeligen Insecten. Zehnter Theil oder vierter Supplementband. H. W. Schmidt, Halle. viii + 320 p. (31 December +)

Macgillavry, D. 1936. Les Bulletins de la Société Entomologique de France. (Bibliographische bijdrage. VII.). Entomol. Ber. 9:204–08. (1 January)

Macquart, P.J.M. 1834. Histoire naturelle des insectes Diptères. Tome première. Ouvrage accompagné de planches. Roret, Paris. 578 p. (before 31 March)
[Published before 31 March teste C. D. Sherborn, unpubl.]

——. 1838. Diptères exotiques nouveau ou peu connus. [Volume 1. Part 1.] Roret, Paris. P. 5–221. (before 17 September)
[This is the reprint version, which was recorded as received in the 17 September issue of C.R. Acad. Sci. Fr. Evenhuis et al. (1989) mention that they had seen a copy of a reprint with the title altered from "Insectes Diptères . . . " to "Diptères exotiques . . . " by a strip of paper with the altered title glued over the original title. Apparently a copy of an unaltered reprint was received by the Academie des Sciences on 17 September 1838, which predates the earliest recorded receipt of the journal version (by Bibliogr. Fr. on 27 October). The altered title (which was continued for the remainder of the series) is maintained here. The journal version was published in 1838, Mém. Soc. R. Sci. Agric. Arts, Lille 1838 (2):9–225. (27 October).]

——. 1840. Diptères exotiques nouveau ou peu connus. Tome deuxième.—1.er partie. Roret, Paris. P. 5–135. (31 December +)
[This is the reprint issued in advance of the journal version published 1841, Mém. Soc. Sci. Agric. Arts, Lille 1840:283–413. (before 7 April)]

——. 1848. Diptères exotiques nouveaux ou peu connus. Suite du 2.me supplément. Mém. Soc. Sci. Agric. Arts, Lille 1847(2):161–237. (before 22 November)
[Presented by Macquart at the 22 November meeting of the Société Entomologique de France. Reprinted with separate pagination, 1–77, Roret, Paris. See also Evenhuis et al. (1989:915) for publishing vagaries in this series.]

——. 1850. Diptères exotiques nouveaux ou peu connus. 4.e supplement [part]. Mém. Soc. Sci. Agric. Arts, Lille 1849:309–479. (before 6 July)
[Recorded in the 6 July issue of Bibliogr. Fr. Reprinted separately, 1850, p. 5–161. Roret, Paris. The reprint version includes only p. 309–465 of the journal version.]

——. 1855. Diptères exotiques nouveau ou peu connus. 5.e supplément. Mém. Soc. Sci. Agric. Arts, Lille (2) 1:25–156. (before 18 August)
[Recorded in the 18 August issue of Bibliogr. Fr. Reprinted with separate pagination, 5–136, Roret, Paris, 1855.]

Malloch, J. R. 1928. Notes on Australian Diptera. No. xvii. Proc. Linn. Soc. N.S.W. 53:598–617. (14 December)

Marschall, A. 1873. Nomenclator zoologicus, continens nomina systematica generum animalium tam viventium quam fossilium, secundum ordinem alphabeticum disposita sub auspiciis et sumptibus C. R. Societatis zoologico-botanicae conscriptus a comite Augusto de Marschall. Vindobonae [= Vienna]. iv + 482 p. (after 20 October)
[Preface dated 20 October.]

Meigen, J. W. 1803. Versuch einer neuen GattungsEintheilung der europäischen zweiflügeligen Insekten. Mag. Insektenkd. 2:259–81. (31 December +)

——. 1804. Klassifikazion und Beschreibung der europäischen zweiflügeligen Insekten (Diptera Linn.). Erster Band. Abt. I. xxviii + 152 p; Abt. II. vi + p. 153–314. Reichard, Braunschweig [= Brunswick]. (late 1804)

[Richards (1935:144), based on research by F. J. Griffin, states that this work was published late in 1804. It therefore postdates Latreille's contributions in vol. 24 of Nouveau Dictionnaire d'Histoire Naturelle.]

——. 1820. Systematische Beschreibung der bekannten europäischen zweiflügeligen Insekten. Zweiter Theil. Forstmann, Aachen. xxxvi + 363 p. (before December)

[Recorded as received in the December issue of Isis (Oken's).]

——. 1822. Systematische Beschreibung der bekannten europäischen zweiflügeligen Insekten. Dritter Theil. Schulz-Wundermann, Aachen. x + 416 p. (before September)

[Recorded as received in the September issue of Isis (Oken's).]

——. 1830. Systematische Beschreibung der bekannten europäischen zweiflügeligen Insekten. Sechster Theil. Schulz, Hamm. xi + 401 + [3] p. (31 December +)

Melander, A. L. 1906. Some new or little known genera of Empididae. Entomol. News 17:370–79. (December)

——. [1928]. Diptera. Fam. Empididae. Genera Insectorum 185. 434 p. (February or March)

[See Arnaud (1966) for dating of this fascicle. Naturae Novitates records receiving fascicules 185–189 simultaneously on 24 April 1928. The Natural History Museum, London, also received fascicules 185–189 simultaneously on 3 May 1928. If these fascicules were mailed simultaneously, then Pierre's fascicle 186 on Tipulidae (Tipulinae) would have to date from 1928, though it has the printed date 1926.]

——. 1950. Taxonomic notes on some smaller Bombyliidae (Diptera) [part]. Pan-Pac. Entomol. 26:145–56. (29 December)

Meunier, F. A. 1915. Nouvelles recherches sur quelques insectes du Sannoisien d'Aix-en-Provence. Bull. Soc. Geol. Fr. (4) 14:187–98. (2 February)

[Though the entire volume 14 of this journal is dated 1914, pages of this issue of the Bulletin have "2 fevrière 1915" printed at the bottom.]

——. 1916. Sur quelques diptères (Bombyliidae, Leptidae, Dolichopodidae et Chironomidae) de l'ambre de la Baltique. Tijdschr. Entomol. 59:274–86. (15 December)

Mik, J. 1882. Litteratur. Diptera. Wien. Entomol. Ztg. 1:44–47. (February)

Morge, G. 1976. Dipteren-Farbtafeln nach den bisher nicht veröffentlichten Original-Handzeichnungen Meigens: "Johann Wilhelm Meigen: Abbildung der europaeischen zweiflügeligen Insecten, nach der Natur." Pars III: Farbtafeln CLXI–CCCV. Beitr. Entomol. 26:543–687. (31 December +)

Mühlenberg, M. 1971. Phylogenetisch-systematische Studien an Bombyliiden (Diptera). Z. Morphol. Ökol. Tiere 70:73–102. (2 August)

Mulsant, E. 1852. Note pour servir a l'histoire des *Anthrax* (insectes dipteres), suivie de la description de trois espèces de ce genre, nouvelles ou peu connues. Mém. Acad. Sci. Belles-Lett., Lyon (Sci.) (N.S.) 2:18–24. (31 December +)

Neave, S. A. 1939a. Nomenclator zoologicus. A list of the names of genera and subgenera in zoology from the tenth edition of Linnaeus 1758 to the end of 1935. In four volumes. Vol. I. A–C. Zoological Society of London, London. xiv + 957 p. (before 20 October**)

——. 1939b. Nomenclator zoologicus. A list of the names of genera and subgenera in zoology from the tenth edition of Linnaeus 1758 to the end of 1935. In four volumes. Vol. II. D–L. Zoological Society of London, London. 1,025 p. (31 December+)

——. 1940a. Nomenclator zoologicus. A list of the names of genera and subgenera in zoology from the tenth edition of Linnaeus 1758 to the end of 1935. In four volumes. Vol. III. M–P. 1,065 p. (19 June*)

——. 1940b. Nomenclator zoologicus. A list of the names of genera and subgenera in zoology from the tenth edition of Linnaeus 1758 to the end of 1935. In four volumes. Vol. IV. Q–Z and supplement. 758 p. (after 30 October)
[Epilogue dated 30 October.]

Newman, E. 1841. Entomological notes [part]. Entomologist 1:220–23. (December)

Oldroyd, H. 1940. Entomological expedition to Abyssinia, 1926–27: Diptera-Brachycera. Tabanidae, Asilidae, Bombyliidae. Ann. Mag. Nat. Hist. (11) 5:192–203. (5 February*)

D'Orbigny, C.V.D., ed. 1845. Dictionnaire universel d'histoire naturelle. . . . Tome sixième. C. Renard, Paris.
[A 13-volume series issued in 12 livraisons per volume. Duponchel, Lucas, and Desmarest are the primary authors responsible for the Diptera entries in this dictionnaire. See Evenhuis (1990b) for full title and dating of the various livraisons and volumes.]

Osten Sacken, C. R. 1877. Western Diptera: Descriptions of new genera and species of Diptera from the region west of the Mississippi, and especially from California. Bull. U.S. Dep. Int., U.S. Geol. Geogr. Surv. Terr. 3:189–354. (30 April)

——. 1886a. Diptera [part]. In: F. D. Godman & O. Salvin, eds., Biologia Centrali-Americana, p. 73–104. Zoologia. Insecta. Diptera. Vol. I. Taylor & Francis, London. (November)
[This series on Diptera was published in parts from July 1886 to August 1903. The parts by Osten Sacken dealing with the Bombyliidae were published as follows: p. 73–104 (October 1886); p. 105–28 (December 1886); 129–60 (January 1887); p. 161–76 (February 1887). See also Osten Sacken, 1886b, 1887.]

——. 1886b. Diptera [part]. In: F. D. Godman & O. Salvin, eds., Biologia Centrali-Americana, p. 105–28. Zoologia. Insecta. Diptera. Vol. I. Taylor & Francis, London. (December)

——. 1887. Diptera [part]. In: F. D. Godman & O. Salvin, eds., Biologia Centrali-Americana, p. 129–60. Zoologia. Insecta. Diptera. Vol. I. Taylor & Francis, London. (January)

Painter, R. H. 1930. Notes on some Bombyliidae (Diptera) from the Republic of Honduras. Ann. Entomol. Soc. Am. 23:793–806. (26 December)

——. 1932. A monographic study of the genus *Geron* Meigen as it occurs in the United States (Diptera: Bombyliidae). Trans. Am. Entomol. Soc. 58:139–67. (15 June)

——. 1933a. New subgenera and species of Bombyliidae (Diptera). J. Kans. Entomol. Soc. 6:5–18. (9 January)

——. 1933b. Notes on some Bombyliidae from Panama. Am. Mus. Novit. 642:1–10. (5 July)

——. 1959. A new genus of Bombyliidae (Diptera). J. Kans. Entomol. Soc. 32:73–75. (18 April)

Painter, R. H. & J. C. Hall. 1960. A monograph of the genus *Poecilanthrax* (Diptera: Bombyliidae). Tech. Bull. Kans. St. Univ. Agric. Exp. Stn. 106:1–132. (15 June)

Painter, R. H. & E. M. Painter. 1963. A review of the subfamily Systropinae (Diptera: Bombyliidae) in North America. J. Kans. Entomol. Soc. 36:278–348. (19 November)

——. 1965. Family Bombyliidae. In: A. Stone, C. W. Sabrosky, W. W. Wirth, R. H. Foote & J. R. Coulson, eds., A catalog of the Diptera of America north of Mexico, p. 406–46. U.S. Dep. Agric., Agric. Handb. 276:i–iv, 1–1696. (23 August)
[Dated from Sabrosky (1967:115).]

——. 1974. Notes on, and redescriptions of, types of South American Bombyliidae (Diptera) in European and United States Museums. Kans. State Univ. Agric. Exp. Sta. Res. Publ. 168:1–322. (February)

Painter, R. H., E. M. Painter & J. C. Hall. 1978. Family Bombyliidae. In: A catalog of the Diptera of the Americas south of the United States 38:1–92. (20 April)

Paramonov, S. J. [1927]. Dipterologische Fragmente. I. *Karakumia* nov. gen. (Bombyliidarum). Zbirn. Prats. Zool. Muz. 1:77–79. (early 1927)
[Dated from Paramonov 1939b:35.]

——. 1929. Beiträge zur Monographie einiger Bombyliiden-Gattungen. (Diptera). Zbirn. Prats. Zool. Muz. 6:1–161. (31 December +)

——. 1931. Beiträge zur Monographie der Bombyliiden-Gattungen *Amictus, Lyophlaeba,* etc. (Diptera). Zbirn. Prats. Zool. Muz. 11:1–218. (before 28 May)
[Recorded in the 28 May issue of Naturae Novitates.]

——. 1933. Dipterologische Fragmente. XXX. Über einige Bombyliiden-Typen des Leningrader Museums. Zbirn. Prats. Zool. Muz. 12:53–56. (31 December +)

——. 1934a. Ueber einige exotische (hauptsachlich südamerikanische) Bombyliiden (Dipteren). Konowia 13:22–34. (15 March)

——. 1934b. Neue und alte Bombyliiden (Dipteren). Stylops 3:107–11. (15 May)

——. 1936. Über neue und alte *Antonia*-Arten (Bombyl. Dipt). Mitt. Dtsch. Entomol. Ges. 7:27–31. (May)

——. 1939a. Eine neue Anthracinen-Gattung (Bombyliidae, Diptera). Dopov. Akad. Nauk URSR 6:23–30. (31 December +)

——. 1939b. Kritichiye oglyal suchasnikh i vikopnikh rodiv Bombyliidae (Diptera) vs'ogo svitu. Trudy Inst. Zool.-Bot. Kiev 23:23–50. (31 December +)

——. 1940. Nasekomye dvukrylye. Sem. Bombyliidae (podsem. Bombyliinae). Fauna SSSR 9(2):i–ix, 1–414. [In Russian with expanded German summary.] (30 September)

——. 1954. Two new genera of Bombyliidae (Diptera) from the Belgian Congo. Proc. R. Entomol. Soc. Lond. (B) 23:26–28. (27 February)

Pérez d'A., V. 1968. Fechas de aparicion de la Revista Chilena de Historia Natural. Rev. Chil. Entomol. 6:135–36. (23 December)

Perris, E. 1839. Notice sur quelques diptères nouveaux. Ann. Soc. Entomol. Fr. 8:47–57. (June)

Philippi, R. A. 1865. Aufzählung der chilenischen Dipteren. Verh. Zool.-Bot. Ges. Wien 15:595–782. (31 December +)

Rafinesque, C. S. 1815. Analyse de la nature ou tableau de l'univers et des corps organisés. Palerme [= Palermo]. 224 p. (between April and 21 July)
[Dated from Richmond (1909:38) and Stafleu & Cowan (1983:557). Richmond (1909:38) states that Rafinesque is thought to have stated that the manuscript for the "Analyse" was completed in April. Rafinesque sailed to America on 21 July and, in all probability, had a few copies (those remaining today) sent before this date. The remainder of the edition was lost in the shipwreck of his vessel on which he had taken passage. Cowan (1970:20) states that Rafinesque lost the stock of this book upon meeting Napoleon's army on his way from Palermo to Paris.]

Richards, O. W. 1935. Notes on the nomenclature of the aculeate Hymenoptera, with special reference to British genera and species. Trans. R. Entomol. Soc. Lond. 83:143–76. (27 June)

Richmond, C. W. 1909. A reprint of the ornithological writings of C. S. Rafinesque. Part I. Auk 26:37–55. (January)

Roberts, F.H.S. 1928. A revision of the Australian Bombyliidae (Diptera). Part i. Proc. Linn. Soc. N.S.W. 53:90–144. (15 May)

——. 1929. A revision of the Australian Bombyliidae (Diptera). III. Proc. Linn. Soc. N.S.W. 54:553–83. (16 December)

Rohdendorf, B. B. 1964. (Historical development of dipterous insects). Akad. Nauk SSSR, Trudy Paleontol. Inst. 100:1–311. (In Russian.) (13 January)
[English translation published 1974. The historical development of the Diptera. Univ. Alberta Press, Edmonton. 360 p.]

Rondani, C. 1856. Dipterologiae italicae prodromus. Vol: I. Genera italica ordinis dipterorum ordinatim disposita et distincta et in familias et stirpes aggregata. A. Stoschi, Parmae [= Parma]. 226 + [2] p. (31 December +)

——. 1861. Dipterologiae italicae prodromus. Vol: IV. Species italicae ordinis dipterorum in genera characteribus definita, ordinatim collectae, methodo analitica distinctae, et novis vel minus cognitis descriptis. Pars tertia. Muscide tachininarum complimentum. A. Stocche, Parmae [= Parma]. 174 p. (31 December +)

——. 1863. Diptera exotica revisa et annotata. Novis nonnullis descriptis. Soliani, Modena. 99 p. (31 December +)
[This is the separate published in advance of Rondani, 1864, which has a slightly different title, but same pagination.]

——. 1864. Dipterorum species et genera aliqua exotica revisa et annotata novis nonnullis descriptis. Arch. Zool. Anat. Fisiol. 3(1):1–99. (May)

——. 1873. Muscaria exotica Musei Civici Januensis observata et distincta. Fragmentum II. Species aliquae in Oriente lectae a March J. Doria, anno 1862–63. Ann. Mus. Civ. Stor. Nat. Genova 4:295–300. (November)

Sabrosky, C. W. 1967. Corrections to A Catalog of the Diptera of America North of Mexico. Bull. Entomol. Soc. Am. 13:115–27. (June)

Sack, P. 1909. Die palaearktischen Spongostylinen. Abh. Senckenb. Naturforsch. Ges. 30:501–48. (31 December+)

Schiner, I. R. 1860a. Vorläufiger Commentar zum dipterologischen Theile der "Fauna austriaca," mit einer näheren Begründung der in derselben aufgenommenen neuen Dipteren-Gattungen. I. Wien. Entomol. Monatschr. 4:47–55. (February)

——. 1860b. In: Fauna Austriaca. Die Fliegen (Diptera). Nach der analytischen Methode bearbeitet von J. Rudolf Schiner. Mit der Charakteristik sämmtlicher europäischer Gattungen, der Beschreibung aller in Deutschland vorkommenden Arten und dem Verzeichnisse der beschriebenen europäischen Arten. I Theil [part], Heft 1, p. 1–80. Gerold, Wien [– Vienna]. lxxxii + 674 p. (31 December+)

——. 1867. Zweiter Bericht über die von der Weltumseglungsreise der k. Fregatte Novara mitgebrachten Dipteren. Section: I. Diptera orthorhapha. Verh. Zool.-Bot. Ges. Wien 17:303–14. (31 December+)

——. 1868. Diptera. In: Reise der Österreichischen Fregatte Novara um die Erde in den Jahren 1857, 1858, 1859, unter den Befehlen des Commodore B. von Wüllerstorf-Urbair. Zoologischer Theil. Band 2, Abtheilung 1, Section B. Wien [= Vienna]. vi + 388 p. (30 April)
[Dated from Higgins (1963:158).]

Schulze, F. E., W. Kükenthal & K. Heider. 1926. Nomenclator animalium generum et subgenerum. Erster Band. A–B nebst Einleitung und Litteraturverzeichnis. Preußischen Akademie der Wissensachaften, Berlin.
[Issued in 5 lieferungen. The initial 3 deal with generic and subgeneric names and have the following dates of issue: p. 1–160 (10 March); p. 161–320 (15 May); p. 321–476 (10 August). The last 2 lieferungen deal with the bibliography. Schulze and Kükenthal died before this volume was completed; Heider was added to complete it.]

Schulze, F. E., W. Kükenthal, K. Heider & R. Hesse. 1927–29. Nomenclator animalium generum et subgenerum. Zweiter Band. C–E. Preußischen Akademie der Wissensachaften, Berlin.
[Issued in 5 lieferungen with the following dates of issue: p. 477–636 (23 May 1927); p. 637–796 (12 October 1927); p. 797–956 (29 February 1928); p. 957–1116 (31 August 1928); p. 1117–1298 (27 March 1929).]

——. 1929–1932. Nomenclator animalium generum et subgenerum. Dritter Band. F–M. Preußischen Akademie der Wissensachaften, Berlin.
[Issued in 6 lieferungen with the following dates of issue: p. 1299–1458 (5 October 1929); p. 1459–1618 (28 March 1930); p. 1619–1778 (7 October 1930); p. 1779–1938 (15 April 1931); p. 1939–2098 (8 October 1931); p. 2099–2184 (3 August 1932).

——. 1932–1935. Nomenclator animalium generum et subgenerum. Vierter Band. N–P. Preußischen Akademie der Wissensachaften, Berlin.
[Issued in 6 lieferungen with the following dates of issue: p. 2185–2248 (3 August 1932); p. 2249–2408 (10 June 1933); p. 2409–2568 (25 November 1933); p. 2569–2728 (3 July

1934); p. 2729–2888 (8 March 1935); p. 2889–2988 (11 December 1935). Heider died before the completion of this volume.]

——. 1936–1954. Nomenclator animalium generum et subgenerum. Fünfter Band. Q–Z. Preußischen Akademie der Wissensachaften, Berlin.
[Issued in 6 lieferungen with the following dates of issue: p. 2989–3054 (before 8 October 1936); p. 3055–3212 (8 October 1936); p. 3213–3372 (3 September 1937); p. 3373–3532 (30 August 1938); p. 3533–3692 (27 January 1940); p. 3693–3716 (26 April 1954).]

Scopoli, J. A. 1763. Entomologica carniolica exhibens insecta carniolae indigena et distributa in ordines, genera, species, varietates. Methodo Linnaeana. Vindobonae [= Vienna]. [30] + 420 p. (31 December +)

Scudder, S. H. 1882a. Nomenclator zoologicus. An alphabetical list of all generic names that have been employed by naturalists for recent and fossil names from the earliest times to the close of the year 1879. I. Supplemental list of genera in zoology. List of generic names employed in zoology and paleontology to the close of the year 1879, chiefly supplemental to those catalogued by Agassiz and Marschall, or indexed in the Zoological Record. Bull. U.S. Natl. Mus. 19[i]:i–xxi, 1–376. (August)
[Recorded in the August issue of Am. J. Sci.]

——. 1882b. Nomenclator zoologicus. An alphabetical list of all generic names that have been employed by naturalists for recent and fossil names from the earliest times to the close of the year 1879. II. Universal index to genera in zoology. Complete list of generic names employed in zoology and paleontology to the close of the year 1879, as contained in the nomenclators of Agassiz, Marschall, and Scudder, and in the Zoological Record. Bull. U.S. Natl. Mus. 19[ii]:1–340. (August)
[Recorded in the August issue of Am. J. Sci.]

Séguy, E. 1926. Faune de France. Vol. 13. Diptères (brachycères) (Stratiomyiidae, Erinnidae, Coenomyiidae, Rhagionidae, Tabanidae, Codidae, Nemestrinidae, Mydaidae, Bombyliidae, Therevidae, Omphralidae). Lechevalier, Paris. 308 p. (31 December +)

——. 1930. Risultati zoologici della missione inviata della R. Società Geografica Italiana per l'esplorazione dell'oasi di Giarabug (1926–1927). Insectes diptères. Ann. Mus. Civ. Stor. Nat. Genova 55:75–93. (27 December)

——. 1932. Diptères nouveaux ou peu connus. Encycl. Entomol. (B) II Dipt. 6:125–32. (14 August)
[For dating of the 11 volumes of this journal see Evenhuis (1989a:209).]

——. 1938. Diptera I. Nematocera et Brachycera. Mission scientifique de l'Omo. Mém. Mus. Natl. Hist. Nat., Paris 8:319–80. (1 February)

Sherborn, C. D. 1902. Index animalium sive index nominum quae ab A.D. MDCCLVIII generibus et speciebus animalium imposita sunt. Sectio prima. A kalendis ianuariis, MDCCLVIII usque ad finem decembris, MDCCC. Cantabrigiae [= Cambridge]. lix + 1,195 p. (December)
[Dated from the English Catalogue of Books for 1902.]

——. 1922–1932. Index animalium sive index nominum quae ab A.D. MDCCLVIII generibus et speciebus animalium imposita sunt. Sectio secunda. A kalendid ianuariis,

MDCCCI usque ad finem decembris, MDCCCL. British Museum, London. cxxxi + 7,056 p.

[Issued in parts as follows: Introduction, bibliography and Index A.–Aff., p. 1–128 (July 1922 [recorded in the July issue of Naturae Novitates]); Index Aff.–Anus., p. 129–384 (before November 1923); Index Anus.–Bail., p. 385–640 (November 1923); Index Bail.–Byzos., p. 641–943 (May 1924); Index C–Ceyl., p. 945–1196 (October 1924); Index Ceyl.–Concolor, p. 1197–1452 (January 1925); Index Concolor–Czizeki, p. 1453–1771 (June 1925); Index Daakar–Dorsalis, p. 1773–2008 (November 1925); Index Dorsalis–Eurystomus, p. 2009–2248 (February 1926); Index Eurystomus–funereus, p. 2249–2568 (June 1926); Index funereus–gyzehensis, p. 2569–2880 (November 1926); Index haani–implicatus, p. 2881–3136 (March 1927); Index implicatus–laminella, p. 3137–3392 (July 1927); Index laminella–lyzzia, p. 3393–3746 (December 1927); Index M-A–melanophysa, p. 3747–3970 (May 1928); Index melanophysa–munchissoni, p. 3971–4194 (August 1928); Index munda–nyx, p. 4195–4450 (November 1928); Index o-nigrum–pallens, p. 4451–4690 (March 1929); Index pallens–Phyllochoreia, p. 4691–4930 (June 1929); Index phyllochroma–Pratincola, p. 4931–5138 (October 1929); Index Pratincola–pyxis, p. 5139–5348 (December 1929); Index quacha–ryzo, p. 5349–5702 (June 1930); Index S littera–serratus, p. 5703–5910 (September 1930); Index serratus–squamosus, p. 5911–6118 (December 1930); Index squamosus–szent, p. 6119–6358 (March 1931); Index T–Trichoscelia, p. 6359–6582 (June 1931); Index Trichoscelis–variegatus, p. 6583–6806 (October 1931); Index variegatus–zizyphinus, p. 6807–7056 (February 1932).]

——. 1932–1933. Index animalium sive index nominum quae ab A.D. MDCCLVIII generibus et speciebus animalium imposita sunt. Sectio secunda. A kalendid ianuariis, MDCCCI usque ad finem decembris, MDCCCL. Epilogue, additions to bibliography, additions and corrections, and index to trivialia. British Museum, London. cxlviii + 1,098 p.

[Issued in parts as follows: Epilogue, additions to bibliography, additions and corrections, index to trivialia under genera, pp. 1–208 (June 1932); Index to trivialia under genera (Atherina–Dia), p. 209–416 (August 1932); Index to trivialia under genera (Diabais (Col.)–Lyurus), p. 417–654 (November 1932); Index to trivialia under genera (Mabouya–Prionops), p. 655–878 (April 1933); Index to trivialia under genera (Prionopus–Zyxomma; addenda acetosae–voltzi), p. 879–1098 (July 1933).]

Sherborn, C. D. & B. B. Woodward. 1899. On the dates of the 'Encyclopédie méthodique': additional note. Proc. Zool. Soc. Lond. 1899:595. (2 May)

Smith, K.G.V., R. W. Crosskey & A. C. Pont. 1980. Bibliography of cited literature. In: R. W. Crosskey, ed., A catalogue of the Diptera of the Afrotropical Region, p. 889–1196. British Museum (Natural History), London. 1,437 p. (10 July)

[Dated from a stamped copy in the BMNH Diptera section.]

Speiser, P. 1920. Zur Kenntnis der Dipteren Orthorrhapha Brachycera. Zool. Jahrb. (Syst.) 43:195–220. (23 June)

Stafleu, F. A. & R. S. Cowan. 1976. Taxonomic literature. A selective guide to botanical publications and collections, with dates, commentaries and types. Vol. I: A–G. Second edition. Regnum vegetable 94. Bohn, Scheltema & Holkema, Utrecht. xl + 1,136 p. (31 December +)

——. 1983. Taxonomic literature. A selective guide to botanical publications and collections with dates, commentaries and types. Vol. IV: P–Sak. Second edition. Regnum vegetable 110. Bohn, Scheltema & Holkema, Utrecht; W. Junk, The Hague, Boston. ix + 1,214 p. (31 December +)

Strand, E. 1928. Miscellanea nomenclatoria zoologica et paleontologica. I–II. Arch. Naturgesch. (A) 92(8):30–75. (July)

Théobald, N. 1937. Note complémentaire sur les insectes fossiles oligocènes des gypses d'Aix-en-Provence. Bull. Soc. Sci. Nancy (N.S.) 1937:157–78. (1 March)

Theodor, O. 1983. The genitalia of Bombyliidae (Diptera). Israel Academy of Sciences and Humanities, Jerusalem. 275 p. (31 December +)

Tonnoir, A. L. 1927. Descriptions of new and remarkable New Zealand Diptera. Rec. Canterbury Mus. 3:101–12. (14 December)

Verrall, G. H. 1889. Bigot's orthography. Wien. Entomol. Ztg. 8:265–70. (25 August)

——. 1909a. British flies. Vol. V. Stratiomyidae and succeeding families of the Diptera Brachycera of great Britain. Gurney & Jackson, London. [vii] + 780 p. (1 January)

——. 1909b. Systematic list of the Palaearctic Diptera Brachycera. Gurney & Jackson, London. 34 p. (1 January)
[Bound at the end of Verrall, 1909a, hence assumed to be published at the same time. There is no indication that it was issued separately on a different date.]

Walker, F. 1852. Diptera (Part III). In: Insecta Saundersiana: or characters of undescribed insects in the collection of Willam Wilson Saunders, Esq., F.R.S., &c., p. 157–252. Vol. 1. Van Voorst, London. 474 p. (before 2 August)
[Presented to the Entomological Society of London on 2 August.]

——. 1871a. List of Diptera collected in Egypt and Arabia, by J. K. Lord, Esq., with descriptions of the species new to science [part]. Entomologist 5:255–63. (March)

——. 1871b. List of Diptera collected in Egypt and Arabia, by J. K. Lord, Esq., with descriptions of the species new to science [part]. Entomologist 5:271–75. (April)

Westwood, J. O. 1835. Insectorum nonnullorum exoticorum (ex ordines Dipterorum) descriptiones. Lond. Edinb. Philos. Mag. (3) 6:447–49. (June)

——. 1840. Order XIII. Diptera Aristotle. (Antliata Fabricius. Halteriptera Clairv.). In: An introduction to the modern classification of insects, p. 125–54. Synopsis of the genera of British insects. Longman, Orme, Brown, Green and Longmans, London. 158 p. (June)
[See Evenhuis et al. (1989:984) for dating of this work.]

White, A. 1914. The Diptera-Brachycera of Tasmania. Part I. Families Leptidae, Stratiomyidae, Nemestrinidae & Cyrtidae. Pap. Proc. R. Soc. Tas. 1914:35–74. (10 September)

——. 1916. The Diptera-Brachycera of Tasmania. Part III. Families Asilidae, Bombylidae, Empidae, Dolichopodidae & Phoridae. Pap. Proc. R. Soc. Tas. 1916:148–266. (30 November)

Whitley, G. 1936. Ichthyological genotypes: some supplementary remarks. Aust. Zool. 8:189–92. (31 December +)

Wiedemann, C.R.W. 1817. Neue Zweiflügler (Diptera Linn.) aus der Gegend um Kiel. Zool. Mag. 1(1):61–86. (before July)
[Listed as received in the July issue of Isis (Oken's).]

——. 1819. Beschreibung neuer Zweiflügler aus Ostindien und Afrika. Zool. Mag. 1(3):1–39. (31 December +)

——. 1820. Munus rectoris in Academia Christiano-Albertina iterum aditurus nova dipterorum genera offert iconibuscque illustrat. Holsator, Kiliae [= Kiel]. viii + 23 p. (before October)

[Reviewed in the October issue of Isis (Oken's). Probably received 1 or 2 months earlier.]

——. 1830. Aussereuropäische zweiflügelige Insekten. Als fortsetzung des Meigenschen Werkes. Zweiter Theil. Schulz, Hamm. xii + 684 p. (before 14 December)

[Recorded as received in the December issue of Isis (Oken's). Dating is from the latest advertisement by the publisher of Isis (Oken's) in the December issue.]

Wiegmann, B. M. 1989. A phylogenetic revision of the family Atelestidae (Diptera: Empidoidea) and its implications for the origin of the cyclorrhaphous Diptera. Master of Science Thesis, Univ. of Maryland, College Park. viii + 205 p. (June)

Williston, S. W. 1895. New Bombyliidae. Kans. Univ. Q. 3:267–69. (1 April)

——. 1896. Manual of the families and genera of North American Diptera. Second ed. Rewritten and enlarged. J. T. Hathaway, New Haven, Connecticut. liv + 167 p. (August)

[Received by J. M. Aldrich from Williston in August 1896 (H. Williams, pers. comm.); first advertisement of its publication appeared in the October issue of Psyche.]

——. 1901. Supplement. In: F. D. Godman & O. Salvin, eds., Biologia Centrali-Americana, p. 273–96. Zoologia. Insecta. Diptera. Vol. I. Taylor & Francis, London. (June)

[This supplement to the series on Diptera by Williston was published in parts from 1900–1901 as follows: p. 217–48 (December 1900); p. 249–64 (22 April*); p. 265–72 (29 May*); p. 273–96 (June); p. 297–328 (19 September*); p. 329–32 (28 December*). Other parts of the supplement are authored separately by Aldrich, Wheeler, and Melander.]

——. 1907. Dipterological notes. J. N.Y. Entomol. Soc. 15:1–2. (7 February)

[The article appeared in the March issue, but author's separates were sent out on 7 February.]

——. 1908. Manual of North American Diptera. Third edition. New Haven, Connecticut. 405 p. (31 December +)

[Preface dated June 1908.]

Woodley, N. E. 1989. Phylogeny and classification of the "orthorrhaphous" Brachycera. [Chapter] 115. In: J. F. McAlpine, ed., Manual of Nearctic Diptera. Vol. 3. Research Branch, Agriculture Canada Monograph No. 32. Ottawa. vi + p. 1333–1581. (August)

Wulp, F. M. van der. 1896. Catalogue of the described Diptera from south Asia. The Hague. vi + 229 p. (before 20 June)

[Presented by Wulp at the 20 June meeting of the Nederlandsche Entomologische Vereeniging.]

——. 1899. Verbeteringen en aanvulligen in den "Catalogue of the described Diptera from South Asia." Tijdschr. Entomol. 42:41–57. (18 August)

Zaitzev, V. F. 1966. [Parasitic flies of the family Bombyliidae (Diptera) in the fauna of Transcaucasia]. Nauka, Moscow & Leningrad. 375 p. [In Russian.] (9 July)

——. 1979. [A new genus of bee flies (Diptera, Bombyliidae) from the Palaearctic]. Trudy Zool. Inst. Akad. Nauk SSSR 88:116–19. [In Russian.] (25 April)

——. 1982a. Microstructure of the labella of the fly proboscis. I. Structure of the pseudotracheal closing apparatus. Entomol. Obozr. 61:517–22. [In Russian.] (14 October*) [English translation, 1982b, Entomol. Rev. 61(3):60–69.]

——. 1984a. Microstructure of the labella of the fly proboscis. II. Pseudotracheal frame. Structure and evolution. Entomol. Obozr. 63:35–39. [In Russian.] (13 April*) [English translation, 1984, Entomol. Rev. 63(1):33–41.]

——. 1984b. Anthophilie und Rüsselabellenstruktur der Dipteren. Verh. SIECC X (Budapest) 1983:169–72. (31 December+)

——. 1986. New species of Cretaceous fossil bee flies and a review of paleontological data on Bombyliidae (Diptera). Entomol. Obozr. 65:815–25. (In Russian.) (13 January*) [English translation, 1987, Entomol. Rev. 66(4):150–60.]

——. 1989. Family Bombyliidae. In: Á. Soós & L. Papp, eds., Catalogue of Palaearctic Diptera, p. 43–169. Vol. 6. Therevidae–Empididae. Akadémiai Kiadó, Budapest. 435 p. (1 March)

Zetterstedt, J. W. 1838. Sectio tertia. Diptera. Dipterologis Scandinaviae amicis et popularibus carissimus. In: J. W. Zetterstedt, Insecta Lapponica, p. 477–868. L. Voss, Lipsiae [= Leipzig]. vi + 1,140 p. (March)
[Issued in parts. Though the completed Diptera portion was recorded in the October issue of Isis (Oken's), p. 386–576 (which include the new genus *Platygaster*) was recorded as received in the March issue of Isis (Oken's).]

——. 1842. Diptera Scandinaviae disposita et descripta. Tomus secundus. Officina Lundbergiana, Lundae [= Lund]. P. 441–894. (October)
[Recorded as received in the October issue of Isis (Oken's).]